IMPROVED TEST BANK
FOR

GEOMETRY

SEEING, DOING, UNDERSTANDING

THIRD EDITION

By
Harold R. Jacobs
with
Donald M. Luepke

W. H. FREEMAN AND COMPANY
New York

ISBN-13: 978-0-7167-7594-2
ISBN-10: 0-7167-7594-8

Printed in the United States of America
Fourth printing

W. H. Freeman and Company
41 Madison Avenue
New York, NY 10010
Houndmills, Basingstoke RG21 6XS, England

www.whfreeman.com

Contents

Introductory Comments

This book consists of chapter tests, a midyear examination, and a final examination that may be used with *Geometry: Seeing, Doing, Understanding,* Third Edition. The chapter tests are designed for an examination period of approximately 45 minutes; the midyear and final are designed for an examination period of approximately 100 minutes. Complete answers for all the tests are in a separate section at the end of this book.

I have found that most classes benefit from frequent short quizzes in addition to chapter tests; quizzes encourage regular study and discourage the temptation to cram at irregular intervals. I have not included any quizzes in this book, however, because I feel that they should be spontaneous and adapted to each class.

GEOMETRY

CHAPTER TESTS

GEOMETRY: Test on Chapter 1

Name______________________________

These figures appeared in the first printed version of the *Elements*.

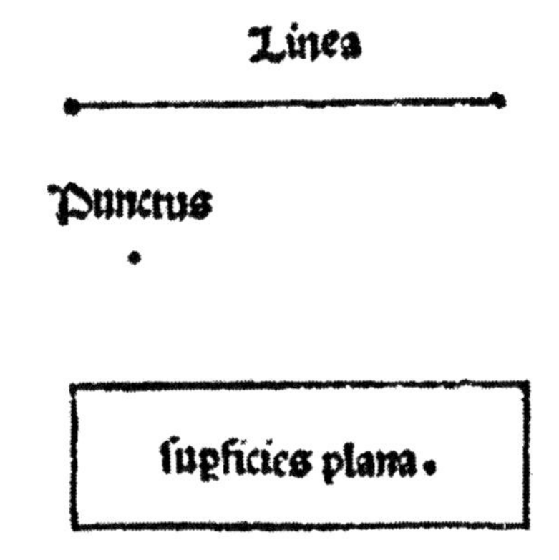

1. Who wrote the *Elements*?
2. What does the figure labeled "punctus" represent?

The figure labeled "linea" looks like a line segment.

3. What is the difference between a *line* and a *line segment*?

The figure labeled "supficies plana" looks like a rectangle.

4. What is the difference between a *rectangle* and a *plane*?

A *quadruped* is an animal with four legs.

5. What is a *quadrilateral*?
6. Explain how the word *pentagon* would help someone figure out how many athletic events are in the *pentathlon*.
7. What do an *octopus* and an *octagon* have in common?

The figure below is a transparent view of a tetrahedron, a polyhedron that has four faces.

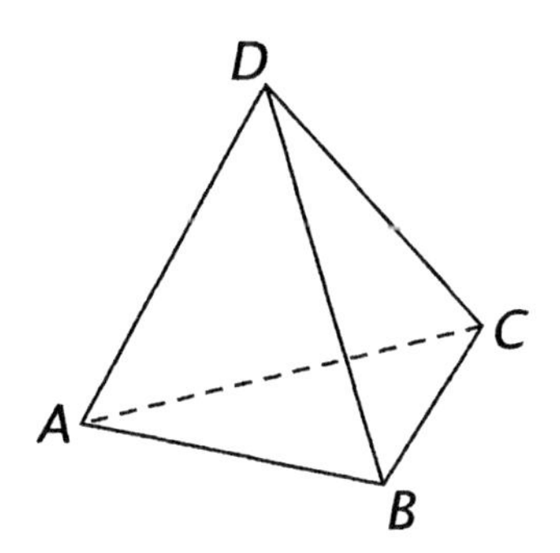

8. What kind of polygon are its faces?
9. How many edges meet at each corner of a tetrahedron?
10. Any three corners of a tetrahedron are noncollinear. What is another word that describes them?

Because of its shape, this quadrilateral is called a "kite."

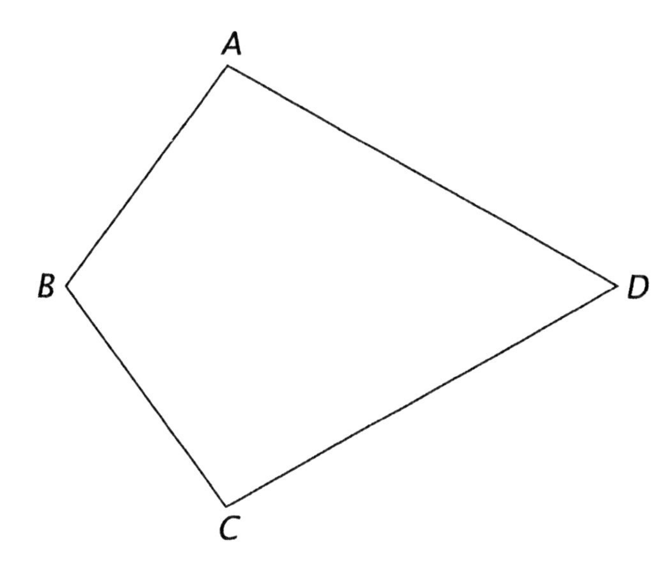

11. Use your straightedge and compass to bisect ∠A, ∠B, and ∠C. Extend the three lines across the figure.
12. What relation do the three lines appear to have to each other?
13. What relation does the line that bisects ∠B appear to have to ∠D?

The floors of two rooms are rectangular in shape. Their dimensions are as follows.

Room A: 15 feet wide, 18 feet long.
Room B: 10 feet wide, 24 feet long.

14. Which room has the greater perimeter? Explain.
15. Which room has the greater area? Explain.

TURN OVER

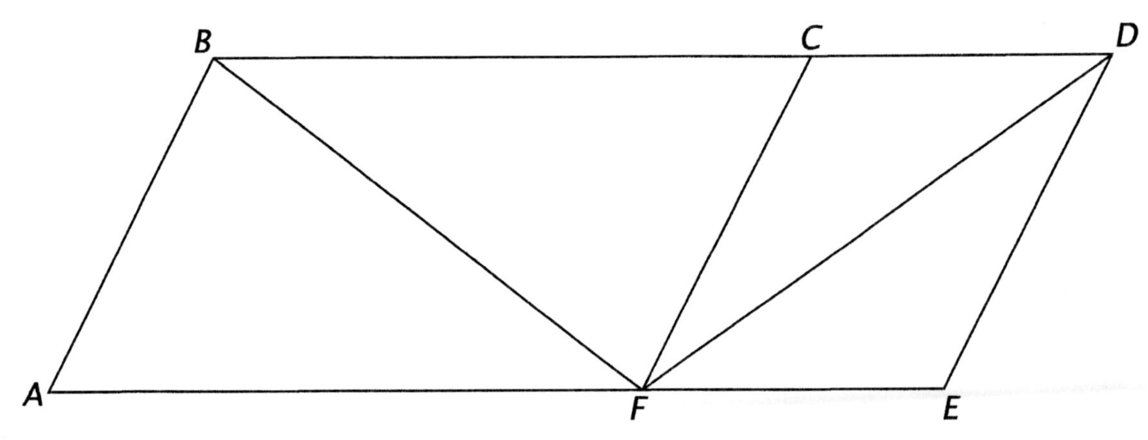

The figure above contains an optical illusion.

Use your ruler to measure the following segments, each to the nearest 0.1 cm.

16. FB.
17. FD.

Use your protractor to measure each of the following angles. (You may extend the sides of the angles as necessary.)

18. ∠A.
19. ∠BFD.

Your measurements reveal something surprising about the figure.

20. What is it?

This figure shows four lines that intersect in six points.

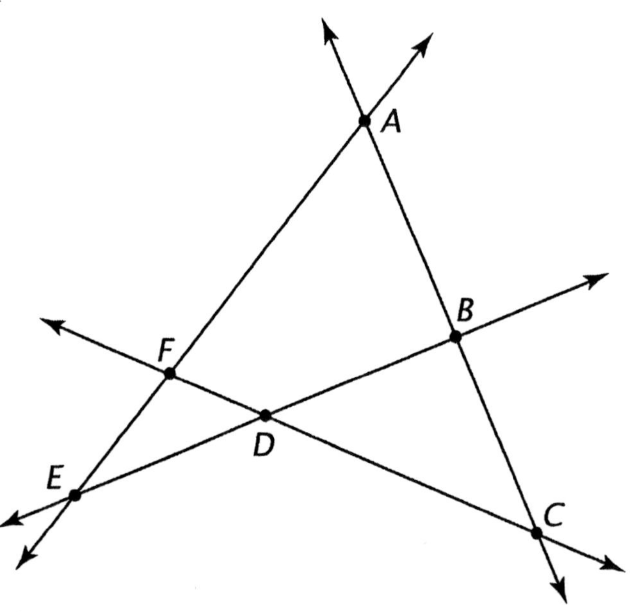

21. Draw AD, BF, and CE. Use your straightedge and compass to bisect these three line segments. Use the letter M to label the midpoint of AD, the letter N for the midpoint of BF, and the letter O for the midpoint of CE.
22. What seems to be true about points M, N, and O?

The figure below shows two quadrilaterals, ABCD and ABEF, drawn on the same grid; AB = DC = FE = 4 and AD = AF = BC = BE = 5.

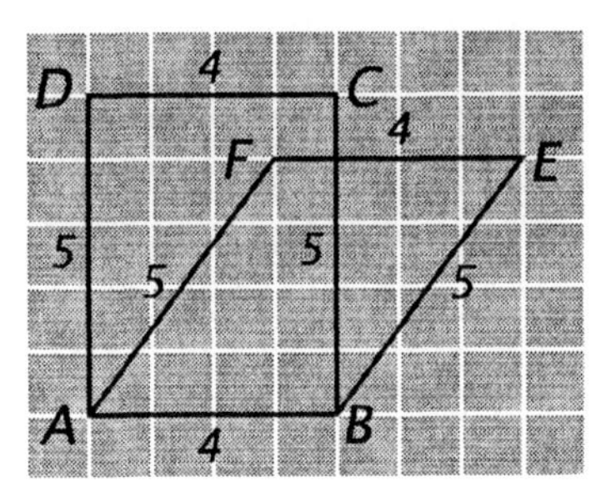

To find the area of a quadrilateral, the Egyptians used the formula

$$A = \frac{1}{4}(a + c)(b + d)$$

in which a, b, c, and d are the lengths of the consecutive sides.

23. Use the figure to show that the Egyptian formula does not always work correctly.

Extra Credit.
A magician takes the three cards shown below from an ordinary deck (containing clubs, ♣, diamonds, ♦, hearts, ♥, and spades, ♠).

1. What cards do you think they are?
2. Can you explain how someone might be tricked with them?

GEOMETRY: Test on Chapter 2

Name______________________________

1. Name the property or definition illustrated by each of the following equations.

   a) $2(3x) = (2 \cdot 3)x$
   b) $4 + y = y + 4$

2. Write each of the following expressions as a single integer.

   a) $(13 - 5)^2$
   b) $13^2 - 5^2$

3. Simplify the following expressions.

   a) $5x^2 - x^2$
   b) $(6x - y) - (x - 7y)$

4. Read the following statements carefully and mark them true or false.

   a) If a conditional statement is true, its converse must also be true.
   b) If the radius of a circle is r, its area is πr^2.
   c) It is possible to define every word in terms of simpler words.
   d) To prove $a \rightarrow b$ indirectly, you begin by assuming *not a*.
   e) A syllogism consists of two premises and a conclusion.

5. This figure illustrates a claim that appeared in a newspaper ad.

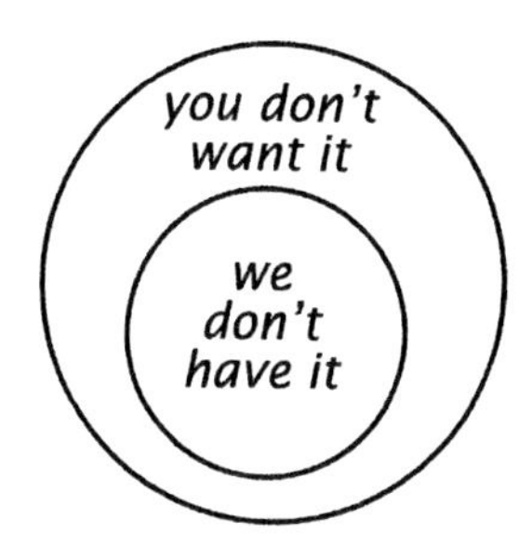

   a) What is this type of figure called?
   b) Write the statement in "if *a*, then *b*" form.
   c) Rewrite it in the form "*b* if *a*."
   d) Which of the following statements is also illustrated by the diagram?
      (1) If you don't want it, we don't have it.
      (2) If you want it, we have it.
      (3) If we have it, you want it.

6. Consider the following premises:

   If you watch Sesame Street, you are a kid at heart.
   If you are a kid at heart, you love Eskimo pies.

   a) What conclusion follows from these premises?
   b) If one of the premises is false, does it follow that the conclusion must be false?
   c) If both premises are true, does it follow that the conclusion must be true?

7. The following statement is a whimsical definition of *egotist:*

   You are an egotist iff you are always me-deep in conversation.

   a) What does the abbreviation "iff" stand for?
   b) Write the two conditional statements that are equivalent to the definition.
   c) How is one of your statements related to the other?

8. The following sentence, from a Spanish geometry book, describes geometry as a deductive system:

   Euclides construye la Geometría partiendo de definiciónes, postulados y axiomas con los cuales demuestra teoremas.

   Write the English equivalent of each of the following words *and* tell what the word means.

   a) definición.
   b) postulado.
   c) teorema.

   d) Tell what you think the sentence says.

9. In this figure of a polyhedron, ABC is a triangle.

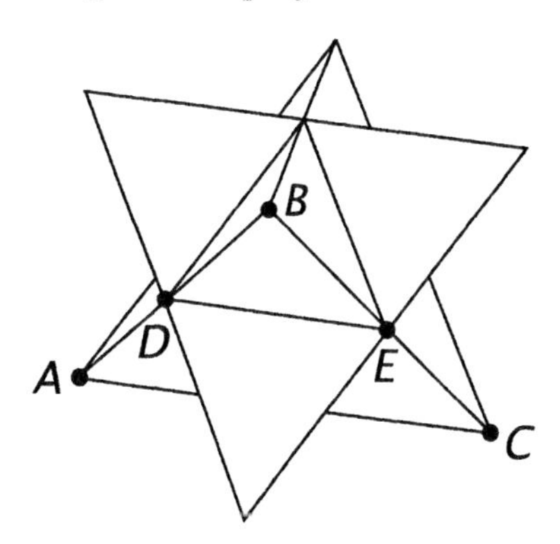

   Tell whether each of the following statements is true or false.

   a) Points A, D, and B are collinear.
   b) Points A and D determine a line.
   c) Points B, E, and C are coplanar.
   d) Points B, E, and C determine a plane.
   e) Points A, D, and C determine a plane.

TURN OVER

10. After studying the relations of the statements given in the following proof, write the missing statements.

Theorem.
If there is a total eclipse of the sun, the temperature can be determined without a thermometer.

Proof.
If there is a total eclipse of the sun, the sky becomes dark.

a) (What is the second statement?)

If the crickets think that it is night, they will start chirping.

b) (What is the fourth statement?)

If the temperature is estimated by counting cricket chirps, it can be determined without a thermometer.

c) What kind of proof is this?

11. Mrs. Cook purchased a set of kitchen utensils advertised as a stainless steel product. After using the set for a few weeks, she discovered that some of the utensils were beginning to rust. She went back to the store, claimed that the set was not stainless steel, and asked for a refund.
In her conversation with the store manager, she used an indirect proof. Identify each of the following.

a) The statement she wanted to prove.
b) The assumption made.
c) The conclusion resulting from the assumption.
d) The known fact contradictory to part c.

12. In this figure, squares have been drawn on the sides of a right triangle. Given that $a = 16$ and $c = 34$, find

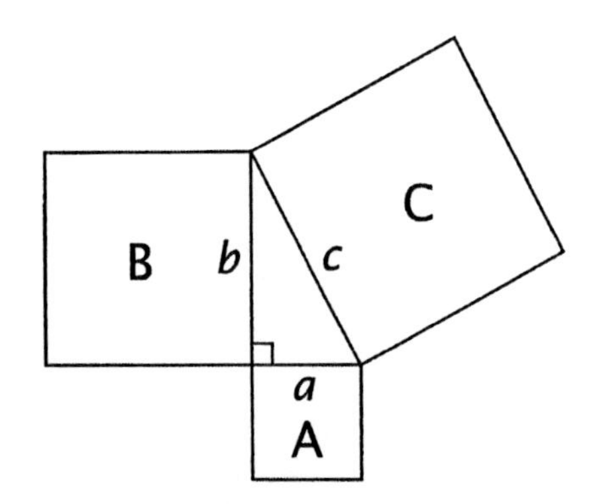

a) the area of square A.
b) the area of square C.
c) the area of square B.
d) b.

13. This figure appeared in a problem on an SAT exam.

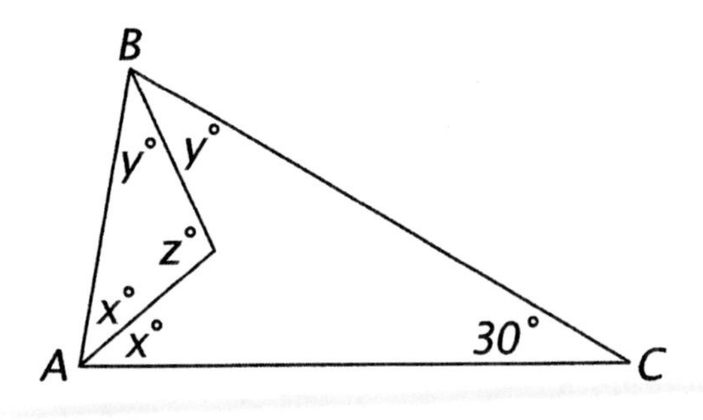

Given that $\angle BAC = 80°$, find each of the following:

a) x.
b) $\angle ABC$.
c) y.
d) z.

14. A trundle wheel can be used to measure distance along the ground. The distance traveled in one revolution of the wheel is equal to the circumference of the wheel.

Given that the circumference of a trundle wheel is 3 feet, find

a) its diameter to the nearest 0.1 inch.
b) the distance that the wheel travels in making five revolutions.

GEOMETRY: Test on Chapter 3

Name______________________________

1. Use the distributive property to eliminate the parentheses.

   a) $9(x - 8)$
   b) $4x(4 + x)$

2. Solve the following equations.

   a) $3x + 20 = 8$
   b) $7(x - 4) = 5x + 2$

3. Read the following statements carefully and mark them true or false.

   a) A line segment has exactly one midpoint.
   b) If two lines intersect, they form two pairs of vertical angles.
   c) Two lines that do not intersect must be parallel.
   d) If two angles are supplementary, then they are a linear pair.
   e) The sides of one of two vertical angles are opposite rays to the sides of the other.

4. The word "complementary" can refer to colors as well as angles, as the following definition shows:

   Two colors are *complementary* iff they add to give white.

   a) Rewrite this sentence in terms of complementary angles by changing as few words as possible.

   The definition of complementary colors refers to colored *light*.

   b) To what kind of angles does the definition of complementary angles refer?

5. This figure shows the design of the flag of Czechoslovakia.

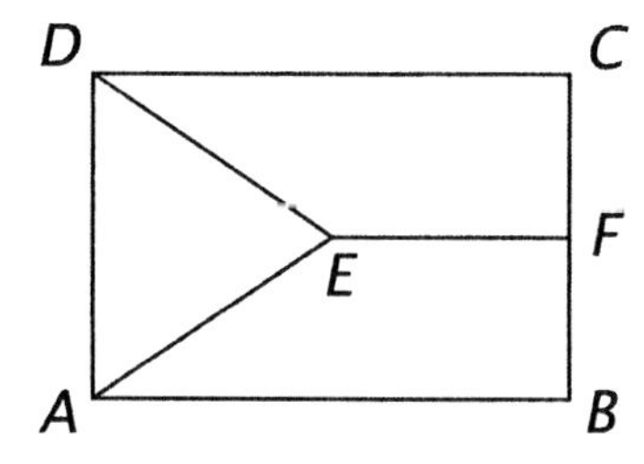

   What type of angle does each of the following angles appear to be?

   a) ∠AEF.   b) ∠AED.   c) ∠EFC.

   Name a pair of angles with a common vertex in the figure that appear to be

   d) complementary.
   e) a linear pair.

6. The figure below is the solution to the following puzzle from a book published in 1821:

   Your aid I want, nine trees to plant
   In rows just half a score;
   And let there be in each row three.
   Solve this: I ask no more.

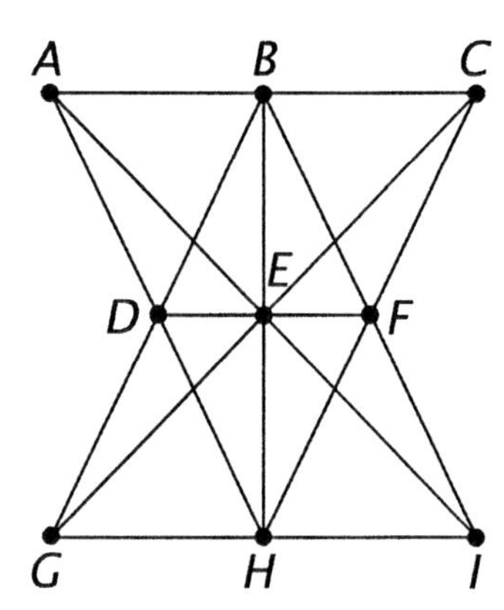

   a) In how many rows are the nine trees planted?

   Write the equation that follows from each of the following statements about the figure.

   b) B-F-I.
   c) HB bisects ∠AHC.
   d) E is the midpoint of GC.
   e) ∠AEB and ∠IEH are vertical angles.

7. This figure represents the line of flight and the shock wave of an airplane flying at three times the speed of sound.

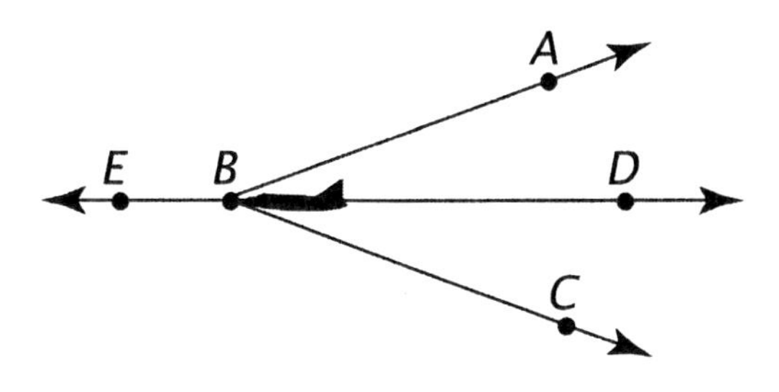

   Line ED bisects ∠ABC, the angle of the shock wave. Given that ∠ABC = 39°, find the measure of

   a) ∠ABD.
   b) ∠ABE.

   If the plane were to fly faster, the angle of the shock wave would get smaller. What would happen to the size of

   c) ∠ABD?
   d) ∠ABE?

TURN OVER

8. This one-foot ruler can be used to measure every whole-number distance from 1 to 12 inches.

A B C D E F

0 2 5 8 11 12

   a) Copy and complete the following statement of the Ruler Postulate:

      The ||||| on a line can be numbered so that positive number ||||| measure |||||.

   b) Of what segment on this ruler is C the midpoint? Explain.
   c) Is point D on this ruler between points C and F? Explain why or why not.

   Name a pair of points that can be used to measure each of the following distances.

   d) 4.
   e) 6.
   f) 9.

9. This figure shows a circular protractor used in navigation. The coordinates of OH and OM are 62 and 114 respectively.

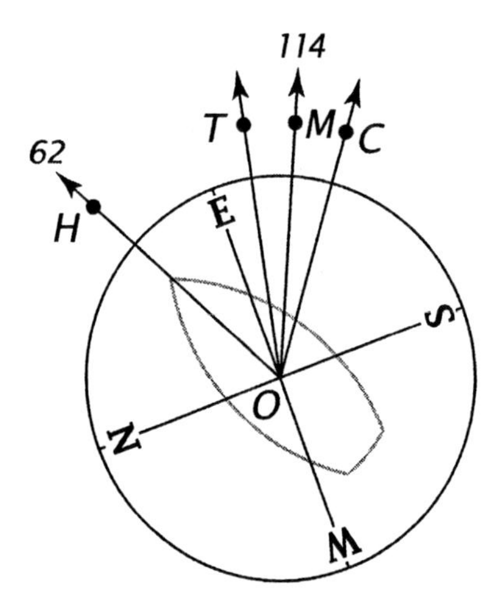

   a) Find ∠HOM.
   b) Find the coordinate of OT given that ∠HOT = 40°.
   c) Find ∠TOM.
   d) Find ∠TOC, given that OM bisects ∠TOC.
   e) Find the coordinate of OC.

10. There are three canals in our inner ears which help us keep our balance. Each is perpendicular to the other two, as illustrated by the three rays in this figure.

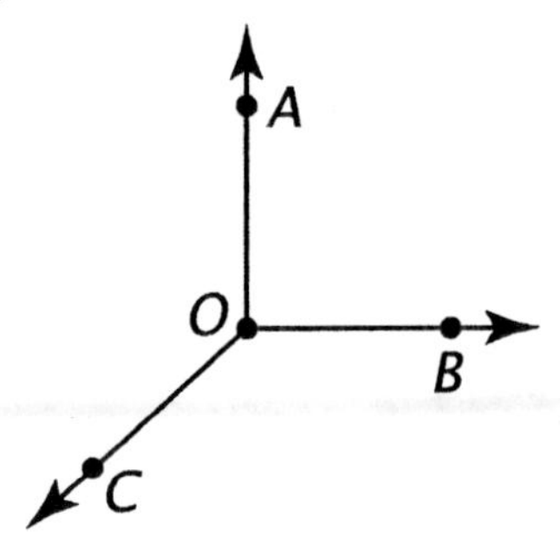

    Explain how we can conclude from this information that ∠AOB, ∠BOC, and ∠AOC are equal.

11. This figure is an overhead view of a magician's magic cabinet. When the mirrored doors BE and CE are closed as shown, a person standing within ΔBEC seems to disappear.

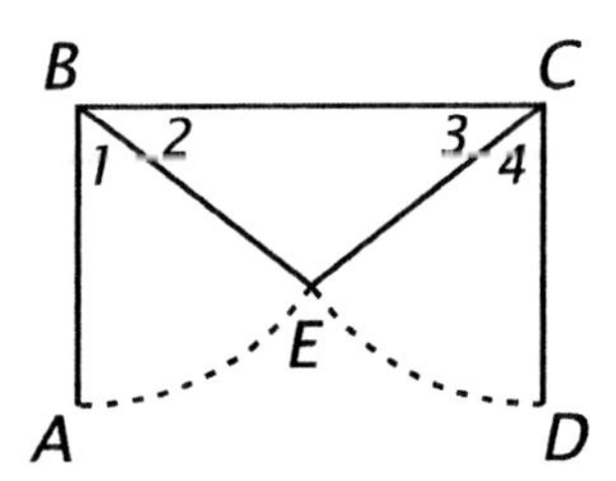

    In the figure, AB ⊥ BC, CD ⊥ BC, ∠2 = n°, and ∠3 = n°.

    Use these facts to explain why ∠BEC = ∠1 + ∠4.

GEOMETRY: Test on Chapter 4

Name____________________

1. Do as indicated.

   a) $(x^2 - x + 7) + (x^2 + 5x - 10)$.
   b) $(9x - 2y) - (x + 3y)$.
   c) $(x + 4)(5x - 2)$.
   d) $(6x + y)^2$.

2. The 5 line segments that form this star meet at their endpoints so that ∠A = ∠B = ∠C = ∠D = ∠E.

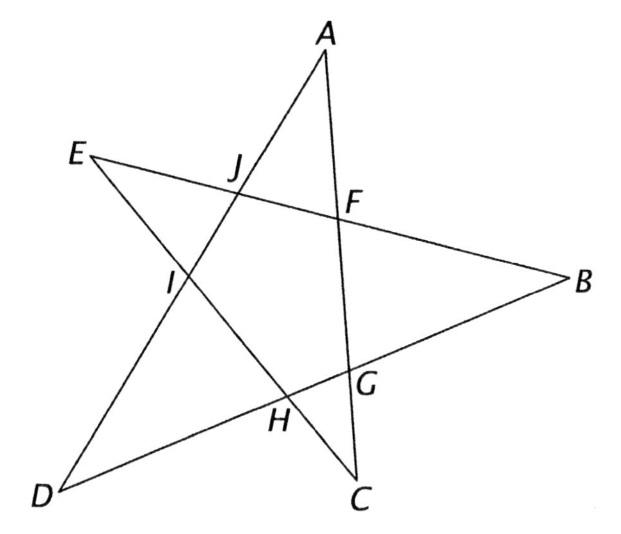

   a) What can you conclude about AG and GD?
   b) State the theorem that is the basis for your answer.

3. These figures show the result of an attempt to copy ΔABC by using a straightedge and compass.

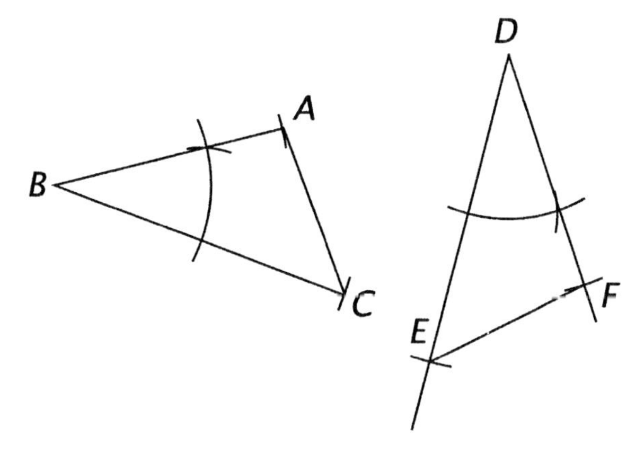

   a) Based on the appearance of the arcs, describe what you think was done.
   b) Is ΔEDF ≅ ΔABC? Explain why or why not.

4. In this figure, AB = AC = AD = AE = BD = CE.

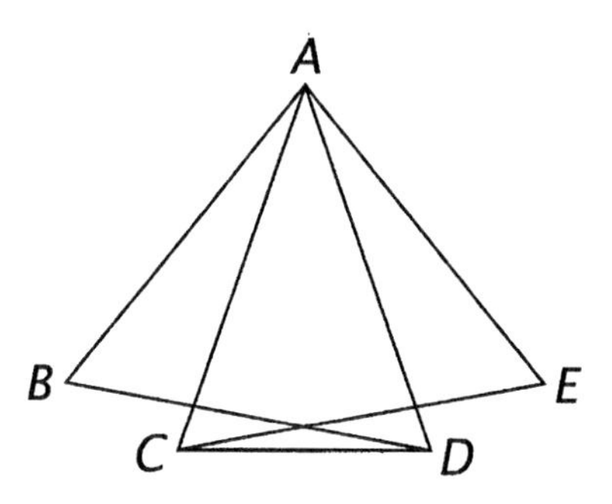

   Tell why each of the following statements must be true.

   a) ΔABD ≅ ΔAEC.
   b) ∠B = ∠E.
   c) ΔABD and ΔAEC are equilateral.
   d) ΔABD and ΔAEC are equiangular.
   e) ΔACD is isosceles.
   f) ∠ACD = ∠ADC.

5. This figure shows positions of the earth, E, and Venus, V, in their orbits around the sun, S. It can be used to estimate the distance from Venus to the sun.

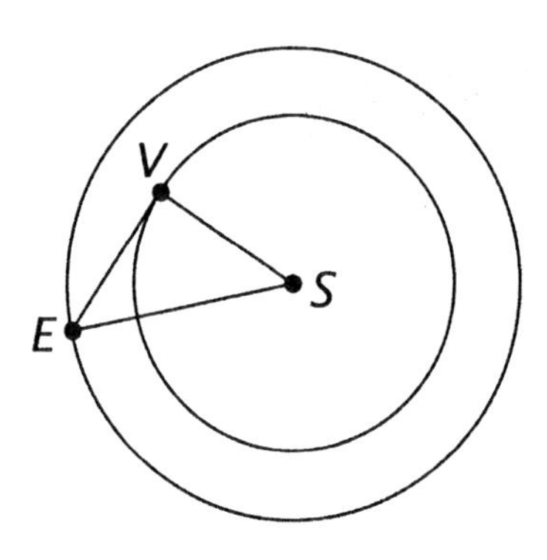

   Given that EV ⊥ VS and ∠E = 45°,

   a) find ∠S.
   b) what can you conclude about ΔVES? Explain.
   c) write an equation relating the lengths of the sides of ΔVES.

   Given also that ES = 93 million miles,

   d) find VS to the nearest million miles.

TURN OVER

6. The first figure below shows a point, P, marked on a rectangular sheet of paper. The second figure shows the paper folded so that the lower left corner, C, falls on P. The third figure shows the paper unfolded flat again.
   Four line segments, PA, PB, PC, and CB have been added to the fourth figure.

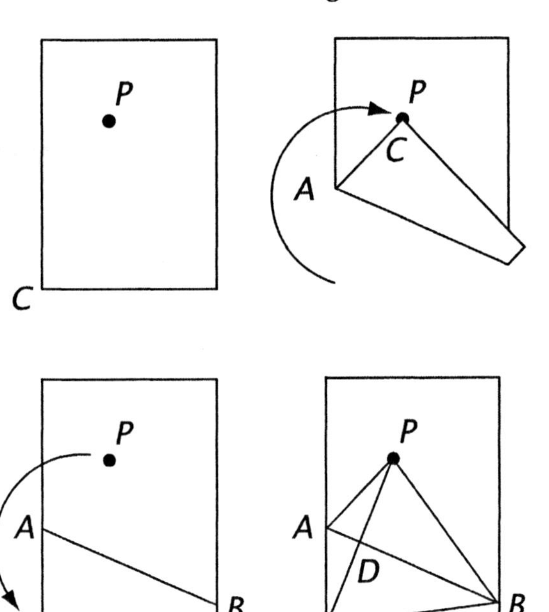

   a) What can you conclude about ΔAPC and ΔBPC?
   b) What can you conclude about ΔAPB and ΔACB?
   c) Why is $\angle PBA = \angle CBA$?
   d) Why is $\Delta PBD \cong \Delta CBD$?
   e) Why is $\angle PDB = \angle CDB$?
   f) Why is $AB \perp PC$?

7. On graph paper, draw a pair of axes extending 5 units in each direction from the origin.
   a) Plot the following points: A(2, 0), B(5, 4), C(–3, 2), D(–2, –1). Use your ruler to draw lines AB and CD across the grid.
   b) Write the formula for the distance between two points, $P_1\ (x_1, y_1)$ and $P_2\ (x_2, y_2)$.

   Use the formula to find the exact distance between

   c) A and B.
   d) C and D.

   e) What are the coordinates of the point that is collinear with A and B and also collinear with C and D?

8. Write a complete proof for the following. Copy the figure and mark the given information on it. Also copy the "given" and "prove" before writing your statements and reasons.

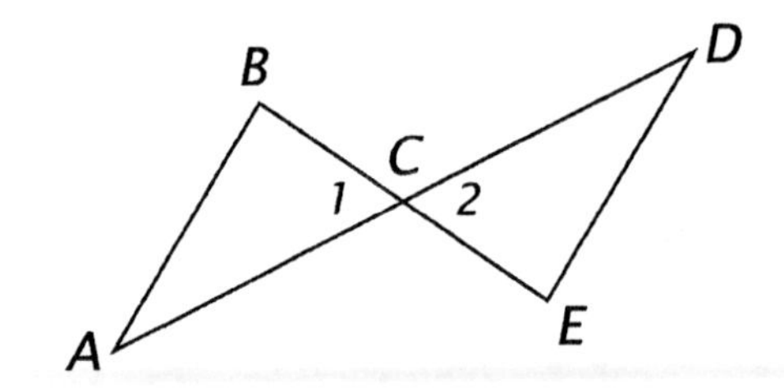

   Given: C is the midpoint of AD; $\angle 1$ and $\angle 2$ are vertical angles; $\angle A = \angle D$.
   Prove: AB = DE.

GEOMETRY: Test on Chapter 5

Name______________________________

1. Factor.

   a) $x^2 + 8x$
   b) $x^2 + 8x + 15$
   c) $x^2 - 9y^2$
   d) $5x^2 + 6x - 8$

2. The figures below represent side views of a deck chair adjusted to two different positions.

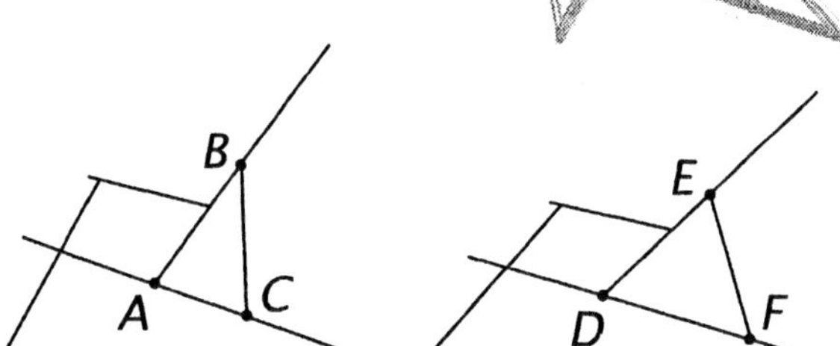

   Tell what symbol should replace the question mark in each of the following statements. (Base your answers on appearances.)

   a) AB = DE and BC = EF, but AC ? DF.
   b) ∠ABC ? ∠DEF.
   c) ∠BAC ? ∠EDF.

3. In this figure, ∠ABD is an exterior angle of ΔBCD.

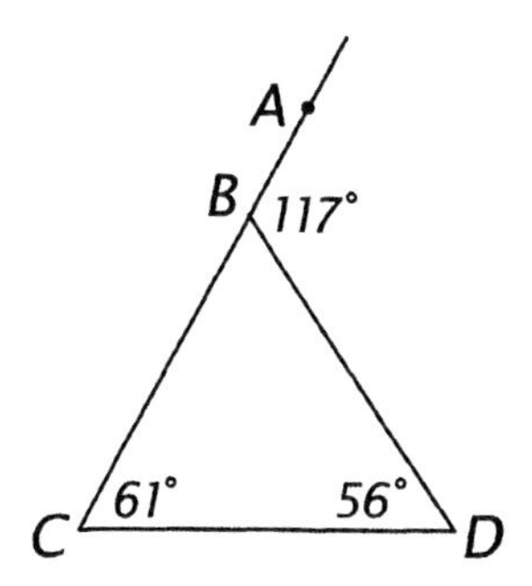

   a) Find ∠CBD.
   b) Which side of ΔBCD is shortest?
   c) Which side of ΔBCD is longest?

4. Read the following statements carefully and mark them true or false.

   a) If $a > b$ and $b > 0$, then $a > 0$.
   b) Every exterior angle of an obtuse triangle is acute.
   c) Each side of a triangle is greater than the sum of the other two sides.
   d) If two angles are unequal, they are not vertical angles.
   e) If a triangle has no equal angles, it must be scalene.

5. AD bisects ∠BAC in ΔBAC.

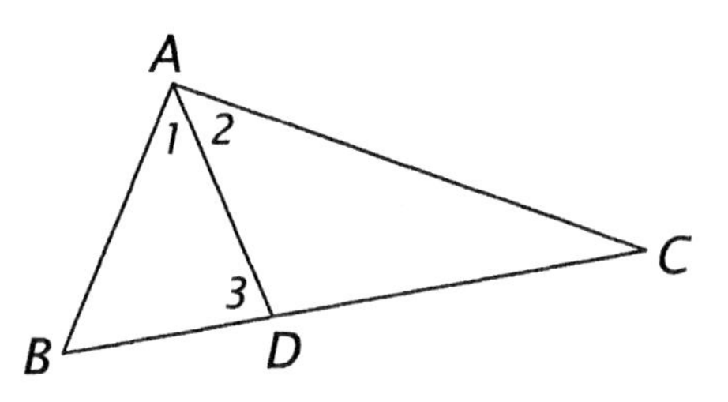

   Prove that $AB > BD$ by giving a reason for each of the following

   a) $\angle 1 = \angle 2$.
   b) $\angle 3 > \angle 2$.
   c) $\angle 3 > \angle 1$.
   d) $AB > BD$.

6. This figure shows one of the strings, *s*, and four of the frets on a guitar.

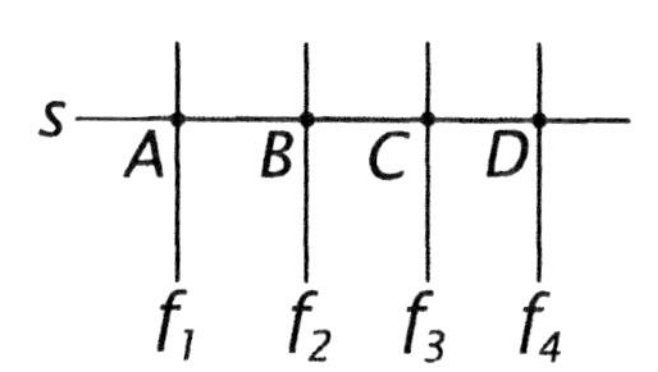

   The frets are spaced under the string so that A, B, C, and D are collinear and $AB > CD$.

   Is it possible to draw any conclusion about the lengths of AC and BD? Explain.

7. The perimeter of this triangle is 12.

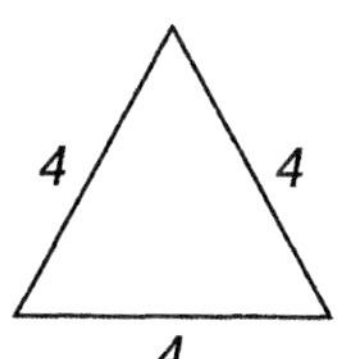

   What other triangles having a perimeter of 12 are possible if their sides also have integer lengths? Show your work.

GEOMETRY: Test on Chapter 6

Name________________________________

1. Reduce to lowest terms.

   a) $\frac{3x}{x^3}$

   b) $\frac{5x-5}{10x-10}$

   c) Express $\frac{1}{4-x}$ as a fraction with denominator $x-4$.

   d) Express as fractions with a common denominator.

   $\frac{2}{x}$ and $\frac{8}{y}$.

2. Read the following statements carefully and mark them true or false.

   a) Through a point not on a line, there is exactly one line parallel to the line.
   b) Each acute angle of an isosceles right triangle is 45°.
   c) In proving a theorem indirectly, we begin by assuming that the hypothesis is false.
   d) In a plane, two lines perpendicular to a third line are perpendicular to each other.
   e) If the hypotenuse and an acute angle of one right triangle are equal to the corresponding parts of another right triangle, the triangles are congruent.

3. Copy and complete the following theorem:

   In a plane, two points each equidistant from the endpoints of a line segment determine . . .

4. The figure below appeared on an SAT exam.

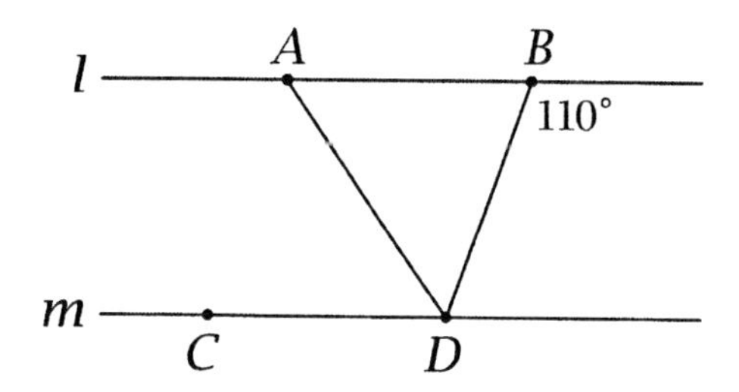

In the figure, $l \parallel m$ and DA bisects ∠CDB.
Find the measure of each of the following angles.

a) ∠CDB.
b) ∠ABD.
c) ∠DAB.

5. The following questions refer to this figure.

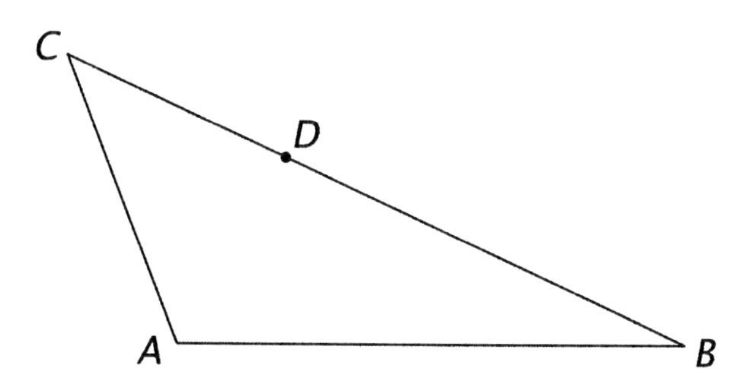

   a) Use your straightedge and compass to construct, through D, a line parallel to CA. Label the point in which the line and AB intersect E.
   b) What relation do ∠A and ∠AED have?
   c) State the theorem that is the basis for your answer.

6. This figure, an illustration in an 18th-century encyclopedia, shows how a cluster of ships might maneuver into a single file.

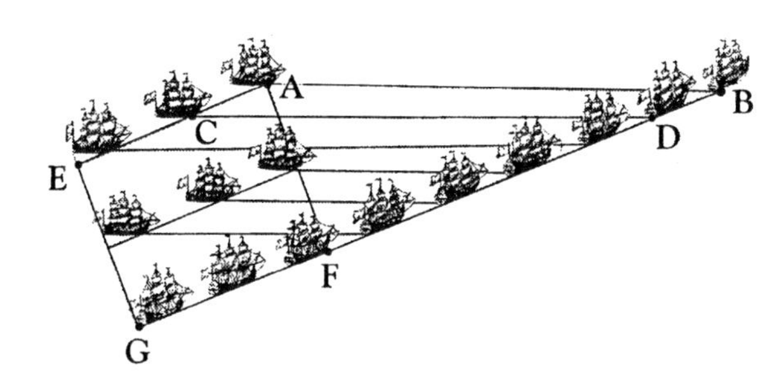

   a) Given that AB ∥ CD, what can you conclude about ∠ABD and ∠CDF?
   b) Why?
   c) Given that AF ⊥ GB, what can you conclude about ∠FAB and ∠B?
   d) Why?
   e) Given that ∠EAF = ∠AFB, what can you conclude about EA and FB?
   f) Why?

TURN OVER

7. In ΔABC, $AD = BE$, $AD \perp BC$ and $BE \perp AC$.

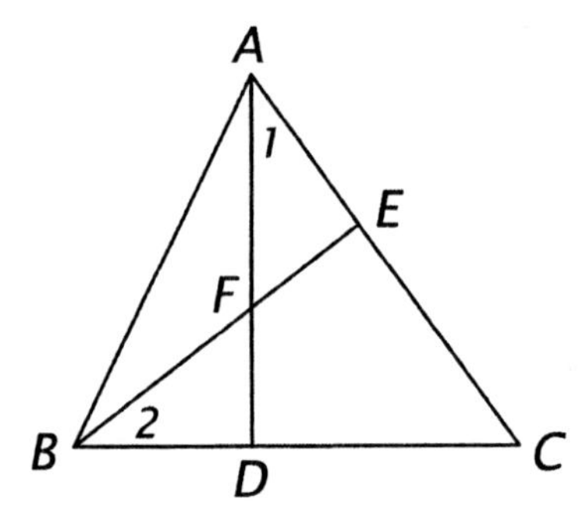

Mark the figure as needed to help in telling why each of the following statements is true.

a) $\angle AFE = \angle BFD$.
b) $\angle 1 = \angle 2$.
c) $\Delta ADC \cong \Delta BEC$.
d) $AC = BC$.
e) $\angle BAC = \angle ABC$.

8. In ΔABC, $AD \perp BC$. Explain why $\angle 1 - \angle 2 = \angle C - \angle B$.

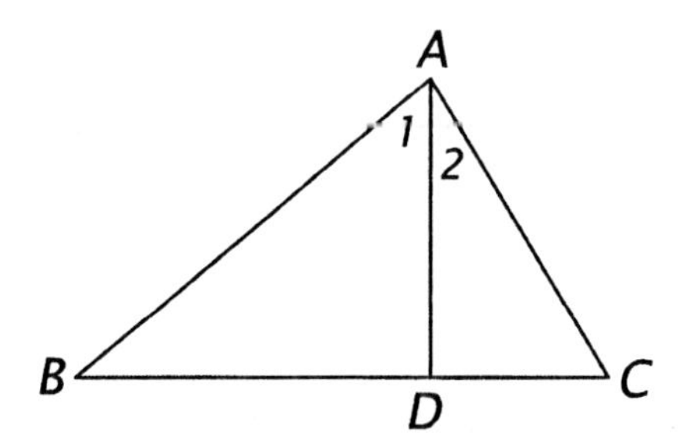

GEOMETRY: Test on Chapter 7

Name________________________________

1. Do the indicated operations. Express your answers in lowest terms.

   a) $\frac{x}{2} + \frac{x}{5}$

   b) $\frac{x+6}{4} \cdot \frac{2}{3}$

   c) $\frac{1}{7} \div \frac{7}{x}$

   d) $\frac{3x+3y}{x^2-y^2} - \frac{1}{x-y}$

2. Copy and complete the following definition.

   A *rhombus* is a quadrilateral ...

3. Read the following statements carefully and mark them true or false.

   a) A parallelogram can have exactly one right angle.
   b) The diagonals of a rhombus bisect each other.
   c) Two consecutive sides of a trapezoid cannot be perpendicular.
   d) A quadrilateral is a parallelogram if it is equiangular.
   e) If the diagonals of a quadrilateral are equal, it is a rectangle.

4. The driving wheels of a steam locomotive are linked by connecting rods so that they turn together.

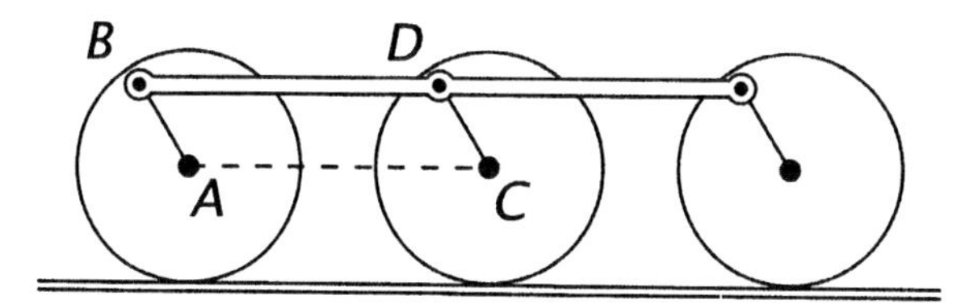

As the wheels turn, AB = CD and AC = BD. Why, for the position shown, can you conclude that

   a) ACDB is a parallelogram?
   b) BD || AC?

5. ABCD is a trapezoid with bases AB and DC; DP bisects ∠ADC and CP bisects ∠BCD.

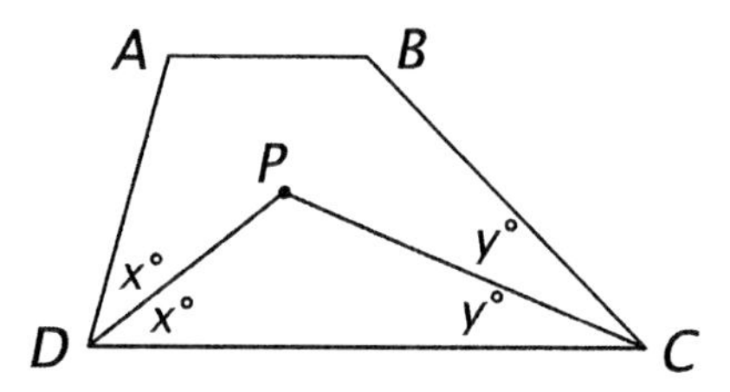

Express each of the following in terms of x and/or y.

   a) ∠P.
   b) ∠A.
   c) ∠B.
   d) ∠A + ∠B.
   e) Write an equation relating ∠P to ∠A + ∠B.

6. An old advertising puzzle consisted of the ten pieces shown below.

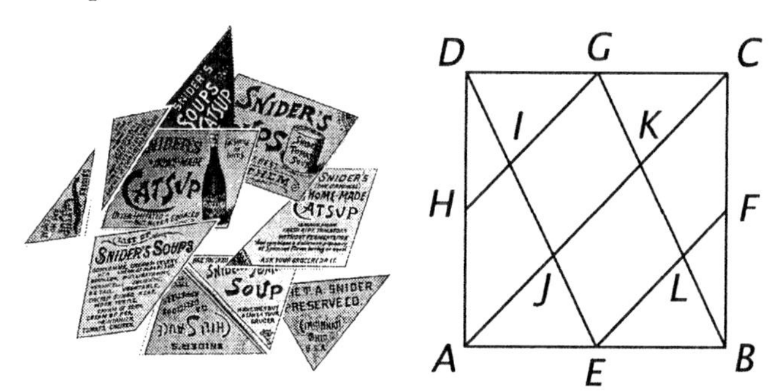

The object was to fit the pieces together to form a square and the solution is shown by the figure to the right of the puzzle. Points E, F, G, and H are the midpoints of the sides of the square ABCD. The figure has been divided by five cuts to form the ten pieces.

   a) What shape do the two larger quadrilateral pieces appear to have?
   b) What shape do the two smaller quadrilateral pieces appear to have?

Prove that your answers are correct by explaining the following.

   c) How do you know that GH || AC and that AC || EF?
   d) How do you know that DE || GB? (Hint: Look at DG and EB.)

GEOMETRY: Test on Chapter 8

Name________________________________

1. Do as indicated.

   a) Write $\sqrt{99}$ in simple radical form.

   b) Simplify: $\sqrt{72} + \sqrt{2}$.

   c) Square and simplify: $(8\sqrt{3})^2$.

   d) Simplify: $\sqrt{17^2 - 15^2}$.

2. Read the following statements carefully and mark them true or false.

   a) An isometry is a transformation that preserves distance and angle measure.
   b) A figure and its image under a translation are congruent.
   c) The reflection of a point cannot be the point itself.
   d) Every trapezoid has reflection symmetry.
   e) All transformations are isometries.

3. Use your straightedge and compass to do the following construction.

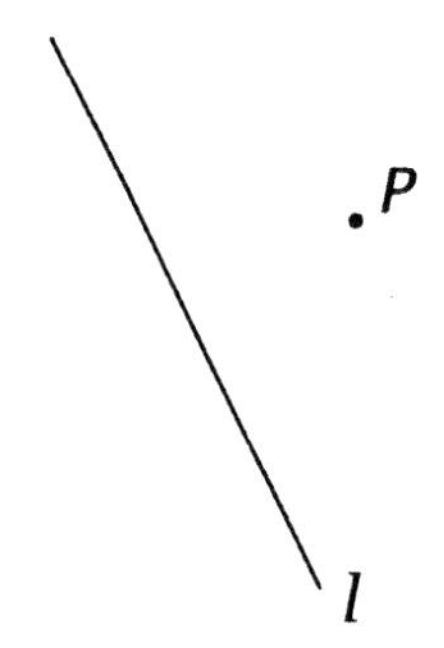

Construct the reflection of point P through line l. Label it P′.

4. This figure shows the pawprints of a bear walking in the snow.

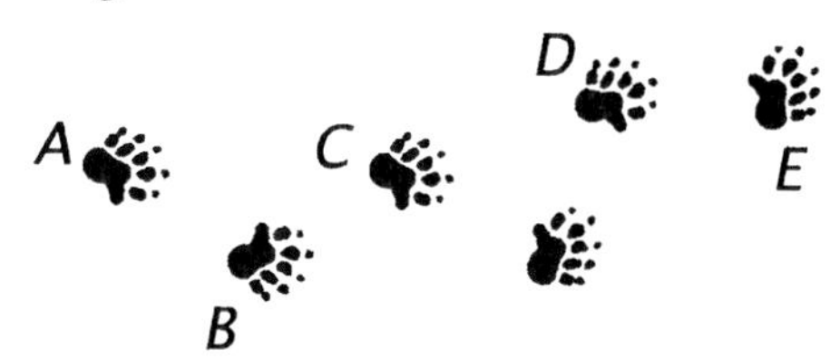

Given that the pawprints are congruent, name the isometry through which each of the following pawprints appears to be the image of A.

   a) B.
   b) C.
   c) D.

   d) Is there an isometry through which E is the image of A?
   e) Explain why or why not.

5. In the figure below, point Z is the image of point X resulting from successive reflections through parallel lines, l_1 and l_2.

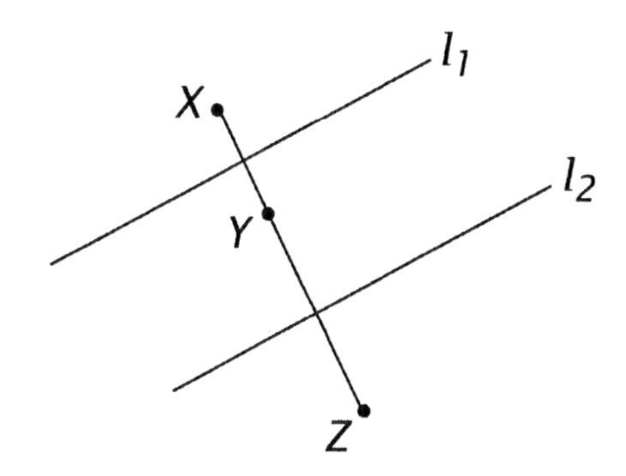

   a) Label the distance from X to l_1 a and the distance from Y to l_2 b. Also label the distances from Y to l_1 and Z to l_2 in terms of a and b.
   b) What is the distance between l_1 and l_2 in terms of a and b?
   c) What is the distance between X and Z in terms of a and b?
   d) Under what transformation is Z the image of point X?
   e) How does the magnitude of the transformation compare to the distance between l_1 and l_2?

TURN OVER

6. A hubcap is shown in the figure below.

   a) What type of symmetry does it have?
   b) To say that the hubcap has "n-fold" symmetry, what number should n be?
   c) What is the measure of the smallest angle through which the hubcap can be turned to look exactly the same?
   d) Would the hubcap look exactly the same if it were turned 216°? Explain.
   e) Does the hubcap have point symmetry? Explain.

7. This figure shows the design of the board on which both checkers and chess are played.

   a) Does the board have any lines of symmetry? If so, where are they?
   b) Does the board have rotation symmetry? Explain why or why not.

8. On graph paper, draw a pair of axes extending 12 units in each direction from the origin.

   a) Plot the following points and connect them to form ΔABC: A(–5, 4), B(6, 1), C(–3, –2).
   a) *(continued)* Plot the following points and connect them to form ΔA′B′C′: A′(–10, 8), B(12, 2), C(–6, –4).
   b) How are the coordinates of the vertices of ΔA′B′C′ related to the corresponding coordinates of the vertices of ΔABC?
   c) For what type of transformation is ΔA′B′C′ the image of ΔABC?
   d) Is this transformation an isometry? Explain.

GEOMETRY: Test on Chapter 9

Name______________________________

1. Read the following statements carefully and mark them true or false.
   a) If one region has a greater perimeter than another, it must also have a greater area.
   b) The perimeter of a rhombus is four times the length of one of its sides.
   c) Parallelograms with equal bases and equal altitudes have equal areas.
   d) Each leg of a right triangle is also one of the triangle's altitudes.
   e) One square foot is equal to twelve square inches.

2. In this figure, the length of each side of the largest square is 24 and the areas of the smaller squares are 256 and 324.

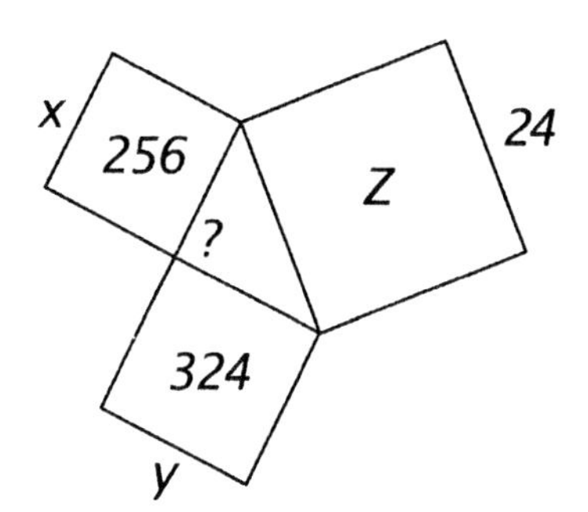

   a) Find length x.
   b) Find length y.
   c) Find area Z.
   d) Is the triangle a right triangle? Explain.

3. This figure appears to contradict part of the Area Postulate.

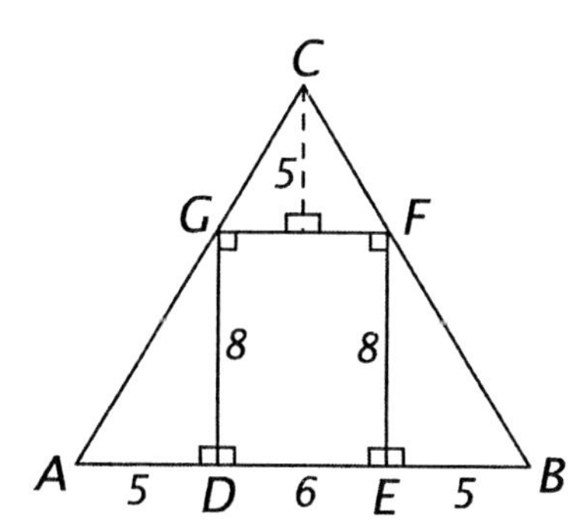

   a) Find the area of the figure by finding the areas of the four parts and adding them.
   b) Find the area of the figure by applying the formula for the area of a triangle to it.
   c) How do these results seem to contradict the Area Postulate?
   d) Assuming that the given lengths are all correct, can you explain the basis for the contradiction?

4. Thirty miles northeast of Mexico City are three ancient pyramids. The region covered by one of them, the Pyramid of the Moon, is shown in the figure at the right below.

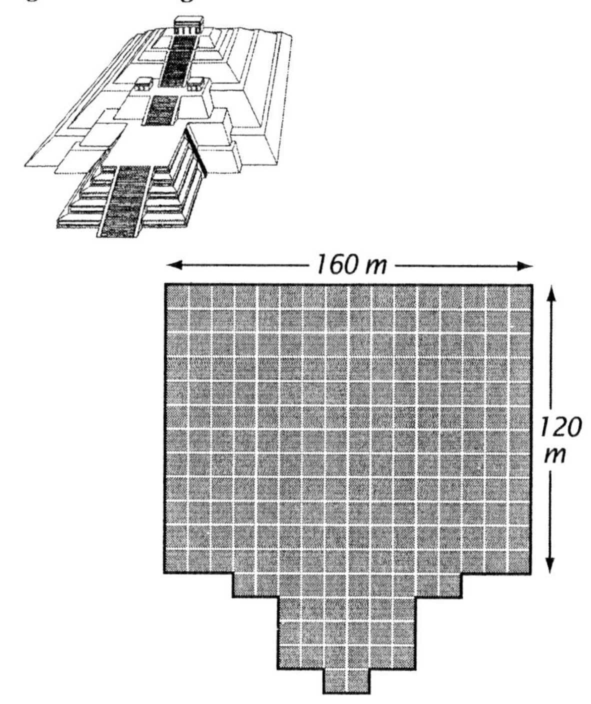

Each small square in the grid has sides of length 10 meters.

Use the figure to find
   a) the perimeter of the region.
   b) its area.

5. This figure, a Maltese Cross, has been drawn on a square grid measuring 5 units on each side.

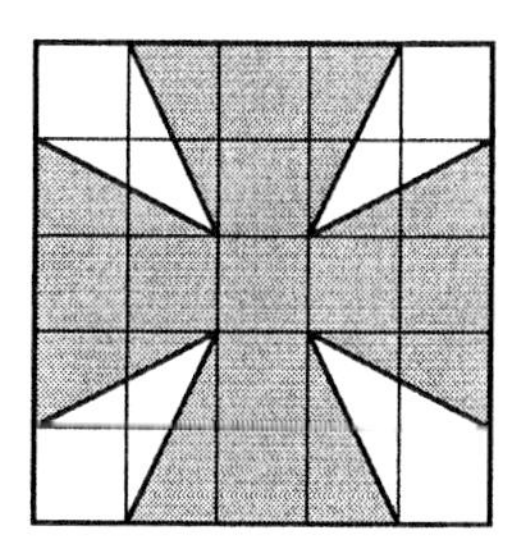

   a) Find the area of the Maltese Cross. Show your method.
   b) Find the area of one of the four kite-shaped quadrilaterals in the corners.

TURN OVER

6. In the first figure below, line segments have been drawn from the vertices of square ABCD to the midpoints of its sides to form a smaller square EFGH.

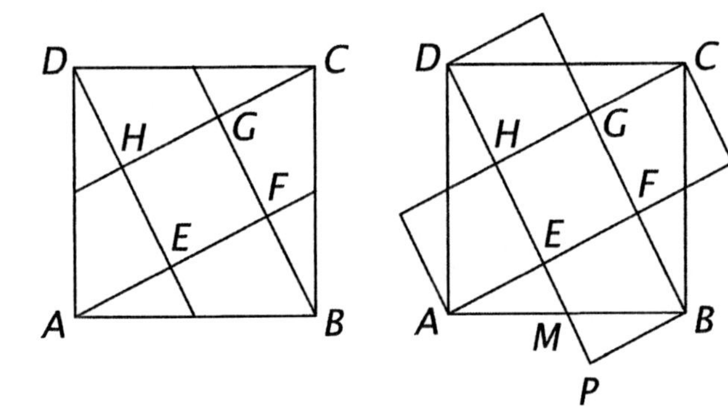

In the second figure, extra line segments have been added to form four more squares.

a) Why is AM = MB?
b) Why is ∠AME = ∠BMP?
c) Why is ∠AEM = ∠BPM?
d) Why is ΔAEM ≅ ΔBPM?
e) Why is αΔAEM = αΔBPM?
f) What does this reveal about the relative areas of square EFGH and square ABCD?

7. This figure appeared in a problem on an SAT exam.

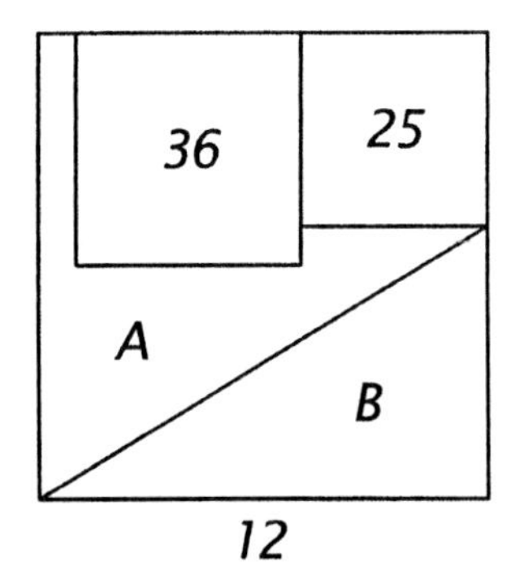

The figure, a square with sides of length 12, is divided into four regions, two of which are squares with areas 36 and 25.

How do the regions labeled A and B compare in area? Explain the basis for your answer.

GEOMETRY: Test on Chapter 10

Name________________________

1. Solve the following equations.

   a) $\frac{12}{x} = 30$

   b) $x - 5 = \frac{x}{3}$

   c) $\frac{5}{x+6} = \frac{3}{x}$

   d) $\frac{x}{4} + \frac{x}{8} = 6$

2. Read the following statements carefully and mark them true or false.

   a) If a line parallel to one side of a triangle intersects the other two sides in different points, it cuts off segments proportional to the sides.
   b) Two right triangles are similar if an acute angle of one is equal to an acute angle of the other.
   c) Corresponding altitudes of similar triangles are equal.
   d) If the corresponding angles of two quadrilaterals are equal, then the quadrilaterals must be similar.
   e) A dilation image of a figure can be smaller than the figure.

3. In New York City, a residential building facing a street w feet wide can be h feet high if $\frac{h}{w} = \frac{3}{2}$.

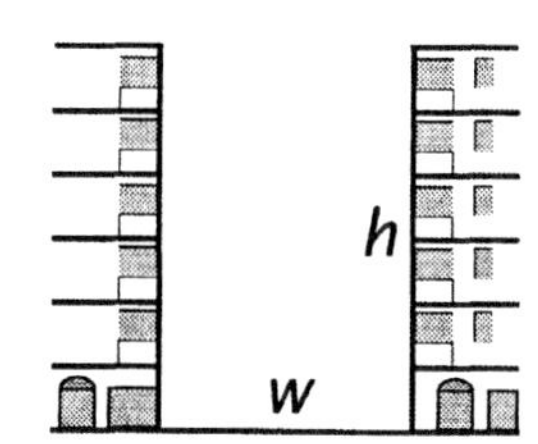

   a) What is an equality between two ratios called?
   b) Solve the formula for h in terms of w.
   c) How high can a residential building be that faces a street 60 feet wide?

4. Write a proportion for each of the following figures and solve each proportion for x.

   a)

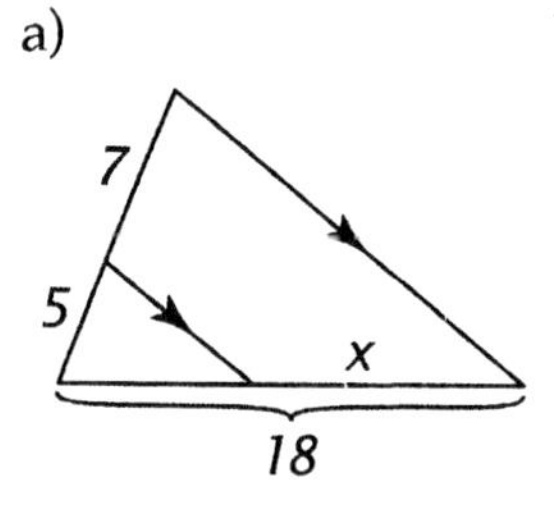

   b)

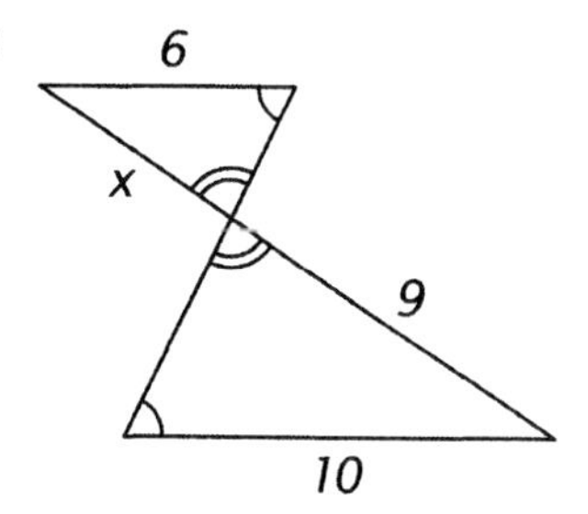

   c)

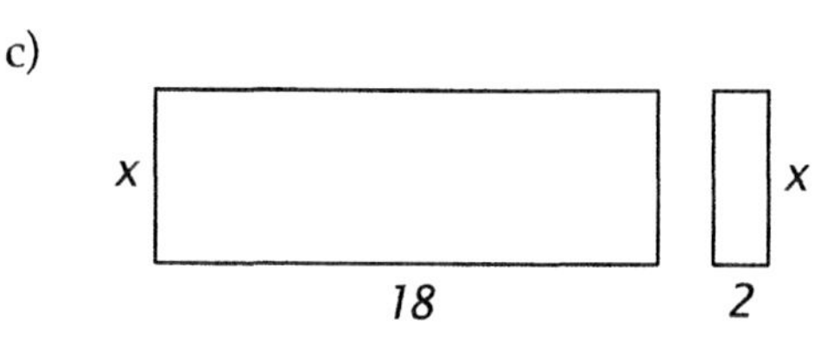

The rectangles are similar

5. These figures, made from tangrams, are similar. The numbers show that the ratio of their corresponding sides is $\frac{3}{5}$ or 0.6.

What is the ratio of their
a) perimeters?
b) areas?

c) Find the perimeter of the second figure given that the perimeter of the first is 45.
d) Find the area of the second figure given that the area of the first is 36.

TURN OVER

6. This problem is from a Chinese book written around 200 B.C.

   A city is 200 pu square. At the center of each side is a gate. At a distance of 15 pu outside the east gate is a tree. Going out the south gate, how many pu does one walk to see the tree?

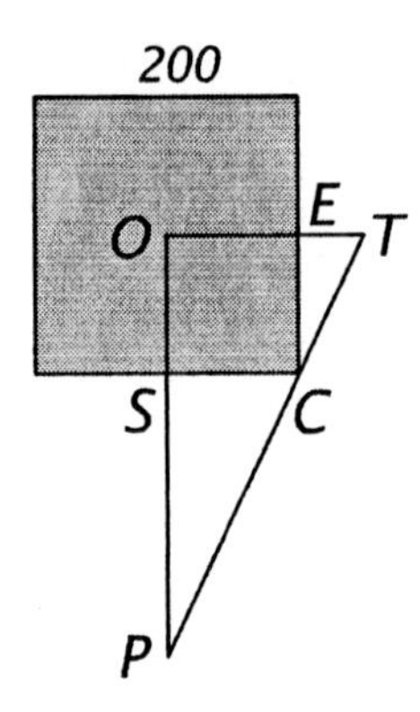

   a) Given that OT || SC and OP || EC, why is $\frac{PS}{SO} = \frac{PC}{CT}$ and $\frac{OE}{ET} = \frac{PC}{CT}$?

   b) Why is $\frac{PS}{SO} = \frac{OE}{ET}$?

   c) Use this proportion and the numbers given in the problem to solve for PS.

7. In this figure, $\Delta CAD \sim \Delta ADB$.

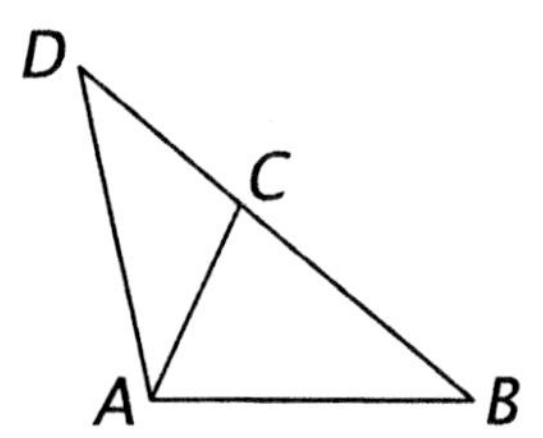

   Mark the figure as needed to help in answering the following.

   a) Why is $\angle CAD = \angle ADB$?
   b) Why is $\angle ADC = \angle DBA$?
   c) What other fact about ΔCAD and ΔADB do these results reveal?

8. In this figure, ΔABC is a right triangle with right $\angle C$ and DEFG is a square.

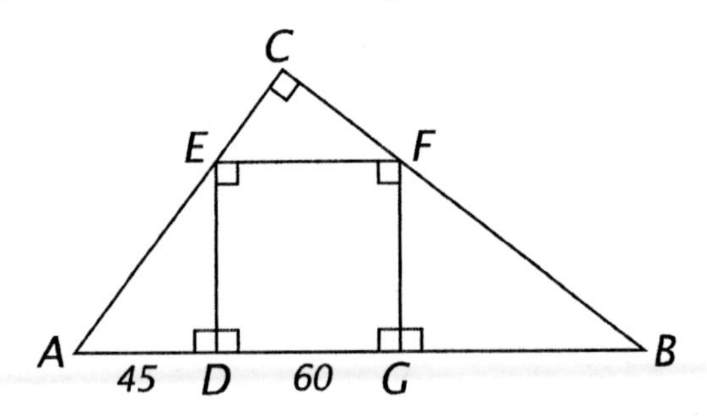

   Given that AD = 45 and DG = 60, find the lengths of the other line segments in the figure.

GEOMETRY: Test on Chapter 11

Name______________________________

1. Solve the following equation.

   a) $x^2 - 12 = 4x$

   Solve the following formulas for the variables indicated. Assume that all of the variables represent positive numbers.

   b) $c^2 = a^2 + b^2$ for b.

   c) $C = \frac{5}{9}(F - 32)$ for F.

2. Find the exact values of the indicated lengths in the following figures.

   a)

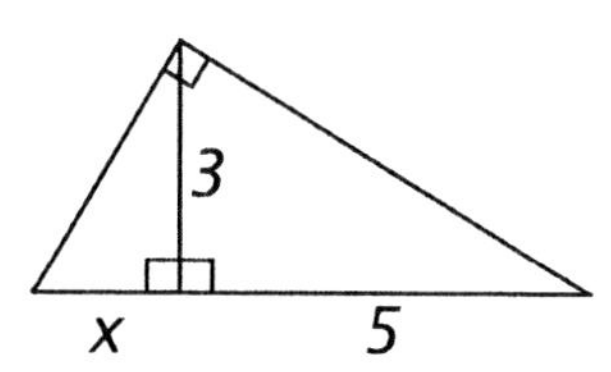

   b)

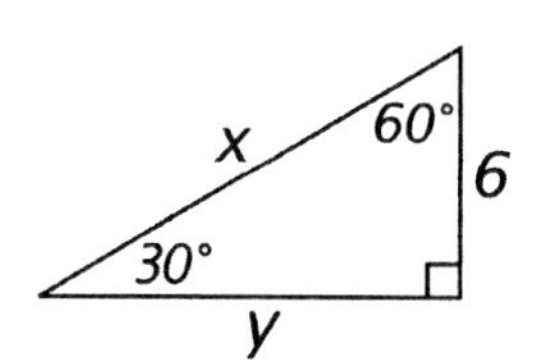

   c)

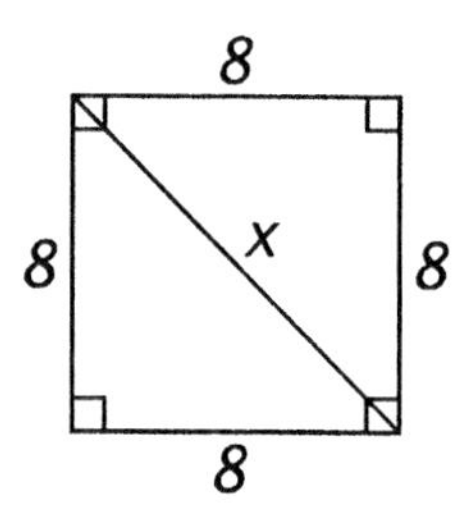

3. This figure was discussed by the Greek philosopher Plato in the 4th century B.C.

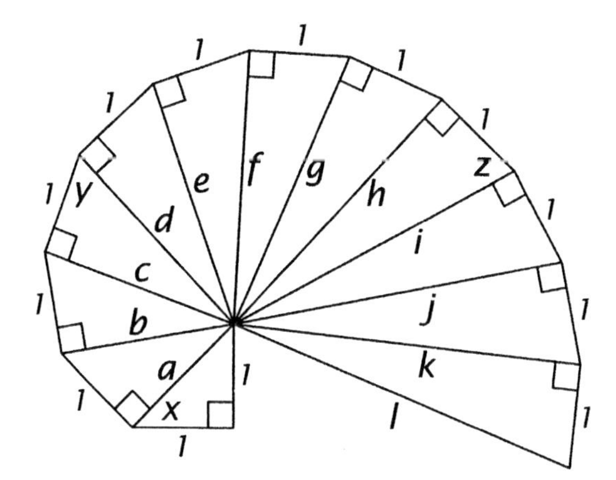

   a) Find the exact lengths of the segments labeled a through l.
   b) Find the measures of the angles labeled x, y, and z, each to the nearest degree.

4. A 100-foot length of copper pipe can expand 0.5 inch when hot water running through it raises its temperature. This has to be taken into account in designing the plumbing of high-rise buildings.

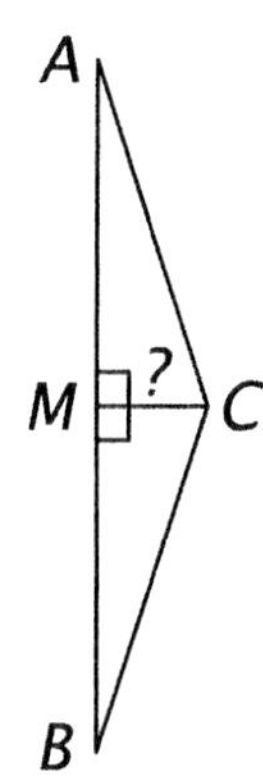

   In the figure above (not drawn to scale), side AB of isosceles ΔABC represents a pipe 100 feet long and M represents its midpoint. Sides AC and CB represent the position of the pipe after it has expanded 0.5 inch.

   a) Label the lengths of AM and AC in inches.
   b) How far does the pipe bend in the middle?

5. This figure, used in trigonometry, shows an angle in a coordinate system.

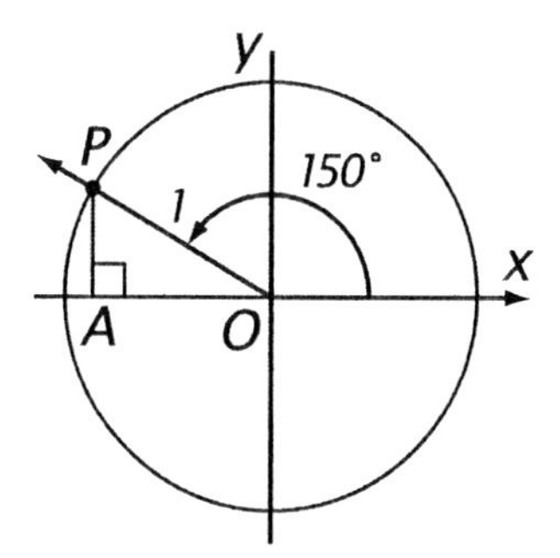

   a) How large is each angle of ΔPAO?

   Given that OP = 1, find

   b) PA and AO.
   c) the coordinates of point P.

TURN OVER

6. This graph shows the path of a hang glider from the top of a hill to the ground below.

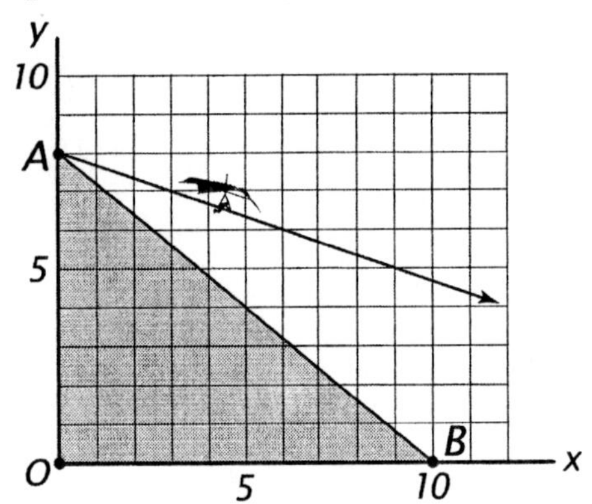

   a) Find the slope of the hill (AB).
   b) Write an equation for the path of the glider in terms of its slope, $-\frac{1}{3}$, and its y-intercept.
   c) Use the equation to find the coordinates of the point in which the glider's path intersects the x-axis (the ground).

7. A pendulum 70 cm long swings between two points 40 cm apart.

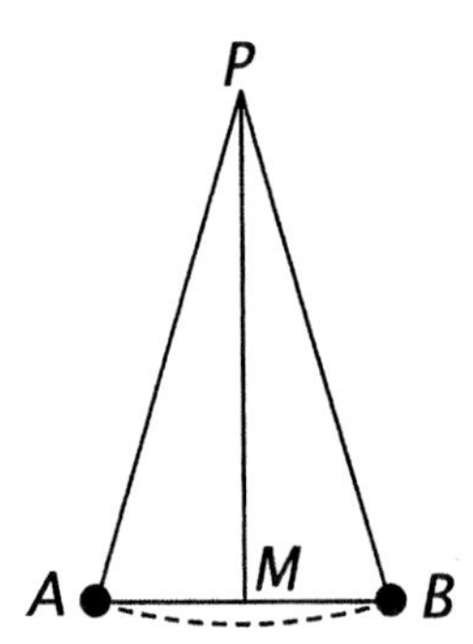

   In this figure, sides PA and PB show the extreme positions of the pendulum and M is the midpoint of AB.

   Use this information to find each of the following.

   a) AM.
   b) ∠APM to the nearest 0.1 degree.
   c) the angle through which the pendulum swings, ∠APB.

8. This figure from a Korean mathematics book illustrates a way to estimate the distance from a boat to positions on the shore.

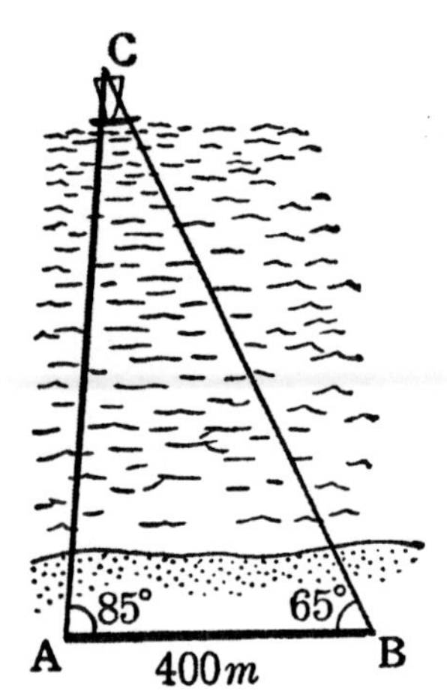

   Use the measurements given on the figure to find the distance from A to C to the nearest meter.

9. The altitude to the hypotenuse of this right triangle with sides of lengths a, b, and c divides it into segments of lengths x and y.

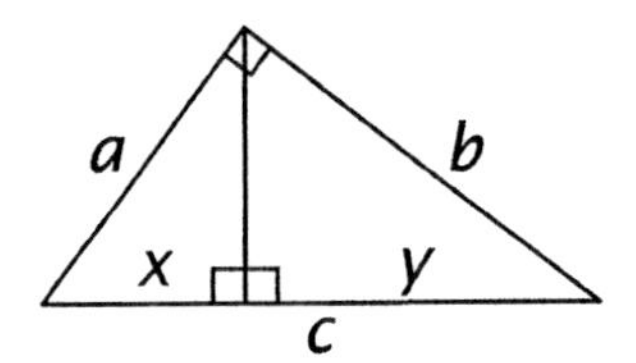

   Explain why $\frac{x}{y} = \frac{a^2}{b^2}$.

GEOMETRY: Test on Chapter 12

Name________________________________

1. A line through the point (3, 4) has a slope of –2.

   a) Sketch its graph on the grid below.

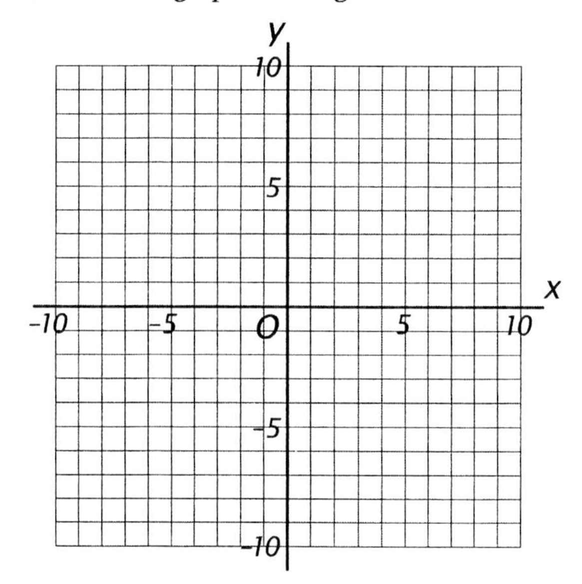

   b) Write its equation in point-slope form.
   c) Transform the equation into slope-intercept form.
   d) What is its *y*-intercept?

2. Read the following statements carefully and mark them true or false.

   a) A line that is perpendicular to a chord also bisects it.
   b) If a line contains the center of a circle, it is a secant of the circle.
   c) If two chords intersect in a circle, they intercept equal arcs.
   d) If a line is perpendicular to a diameter at one of its endpoints, then it is tangent to the circle.
   e) Inscribed angles that intercept equal arcs are equal.

3. Solve for *x* in each of the following figures.

   a)

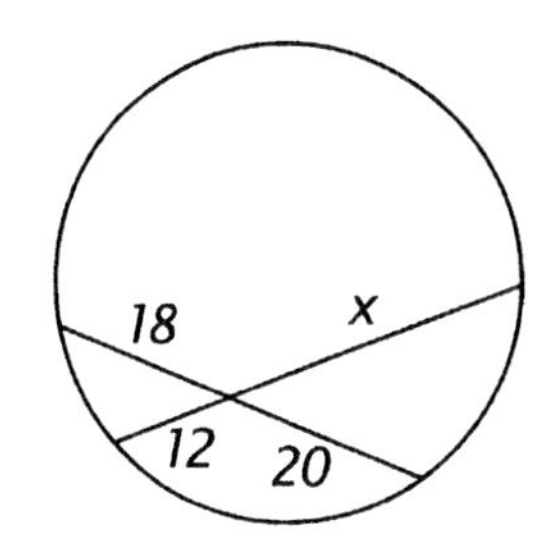

   b)

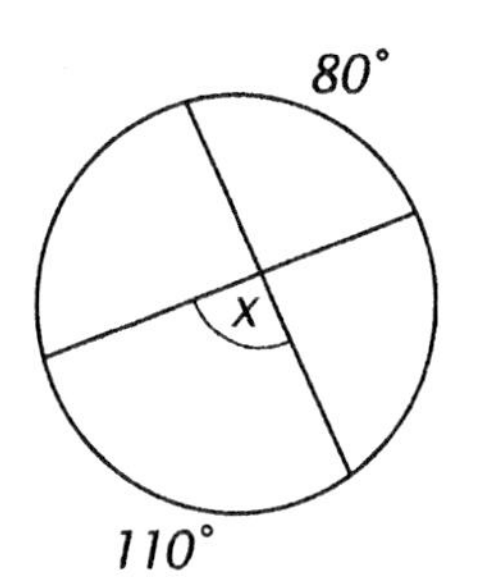

   c)

4. The curves of the runners of a child's rocking horse are arcs of circles.

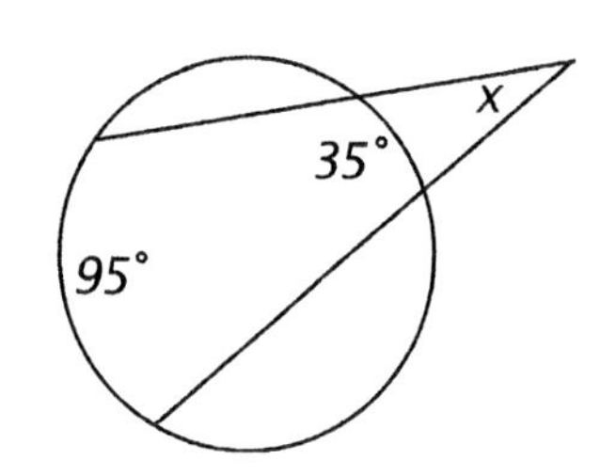

   a) Use your straightedge and compass to find the center of the circle of which the arc above is part.
   b) State the theorem that is the basis for your method.

TURN OVER

5. In the figure below, the points of the star divide the circle into equal arcs.

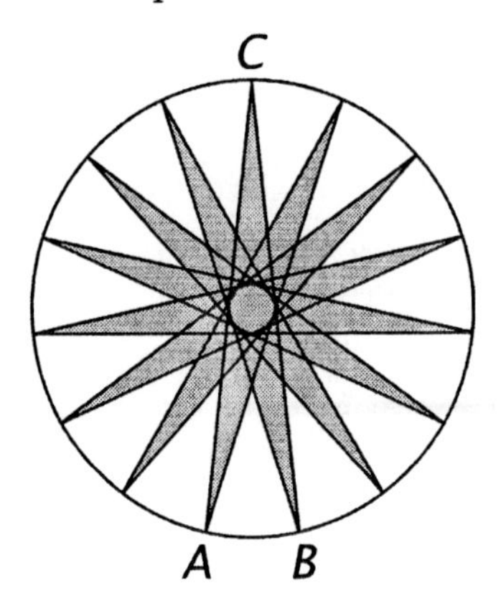

a) Find $m\widehat{AB}$.
b) Find $\angle C$.
c) What is the sum of the 15 angles at the points of the star?

6. The theorem about an angle inscribed in a semicircle was known by the Babylonians as long ago as 2000 B.C.

a) What does the theorem say?

The figure below shows an angle inscribed in *a quarter of a circle.*

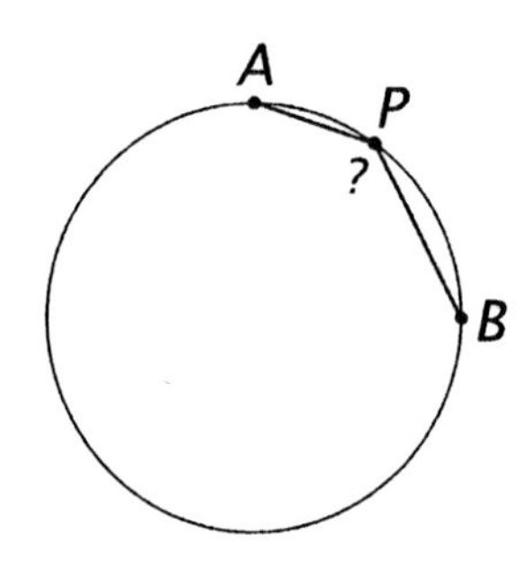

b) What can you conclude about its measure? Explain.

7. The sides of ΔABC are tangent to circle O at X, Y, and Z; AC = 5, CB = 12, and AB = 13.

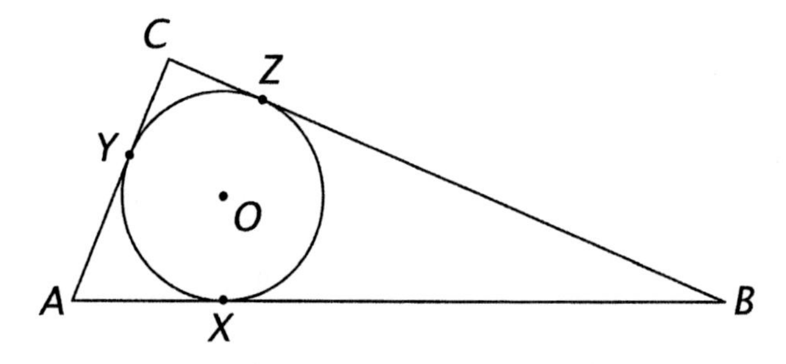

Mark the figure as needed to answer the following questions.

a) What kind of triangle is ΔABC? Explain.

Draw OY and OZ.

b) What kind of quadrilateral is OYCZ?

Given $OY = OZ = r$, label the other lengths in the figure in terms of r.

c) Write an equation and solve it for r.

8. The tallest mountain on Mars is about 15 miles high. Mars has a radius of 2100 miles.

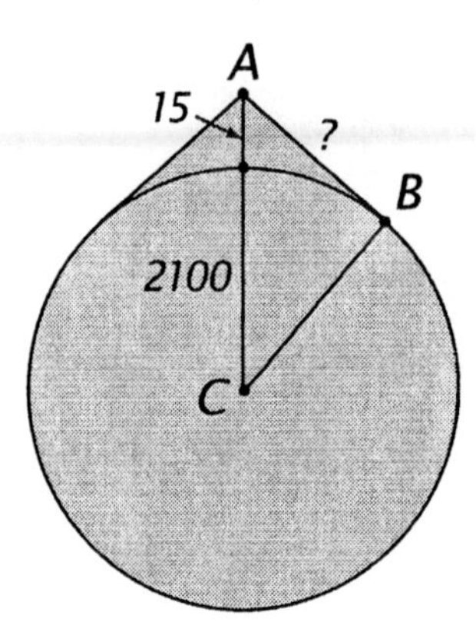

In the figure above (not to scale), point A represents the top of the mountain and AB represents a tangent line of sight to the horizon.

a) Why is $AB \perp BC$?
b) If you stood at point A, approximately how many miles could you see to the horizon (point B)?

9. In this figure, AD and BC intersect at point E in the circle.

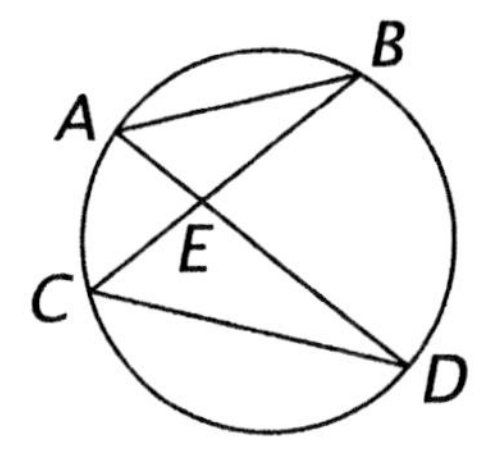

Explain why it follows that $\frac{AB}{CD} = \frac{AE}{CE}$.

GEOMETRY: Test on Chapter 13

Name________________________________

1. Solve the following systems of equations. *Show your methods.*

   a) $x = 3y$
   $5x - 6y = 36$

   b) $4x + y = 18$
   $7x + 5y = 51$

2. Read the following statements carefully and mark them true or false.

   a) The incenter of a triangle is equidistant from its sides.
   b) A circle is inscribed in a polygon if each side of the polygon is tangent to the circle.
   c) The centroid of an obtuse triangle is outside the triangle.
   d) Every rectangle is cyclic.
   e) A median of a triangle is a line segment that connects the midpoints of two of its sides.

3. The three cevians in ΔABC are concurrent.

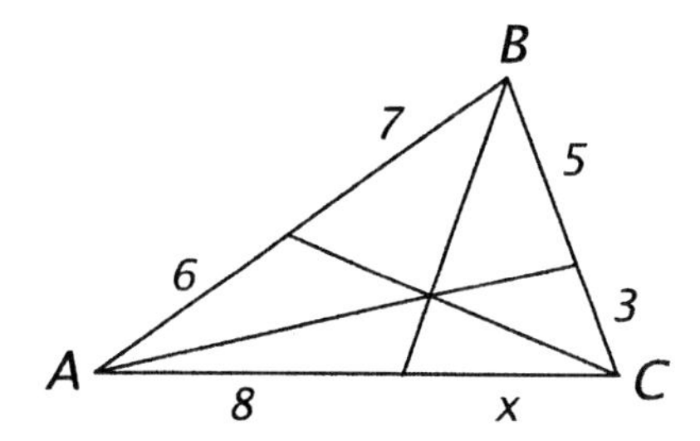

   a) What does the word *concurrent* mean with respect to lines?
   b) Solve for x.

4. Use your straightedge and compass to do the following constructions. Do not erase any arcs or lines used.

   a) Circumscribe a circle about ΔABC.

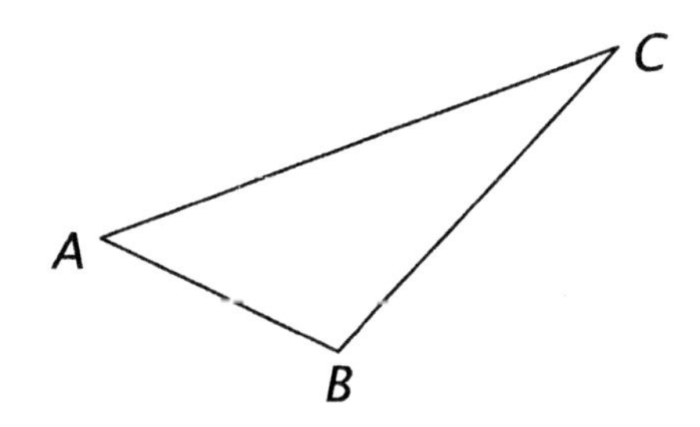

   b) Inscribe a circle in ΔDEF.

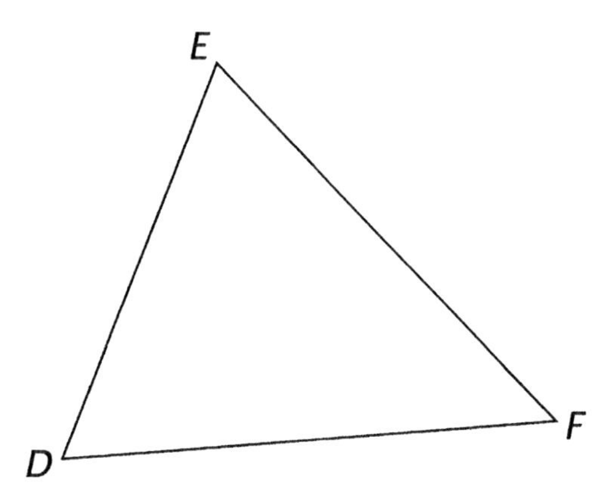

5. Mathematics teacher Dwight Paine wrote a poem about this figure.

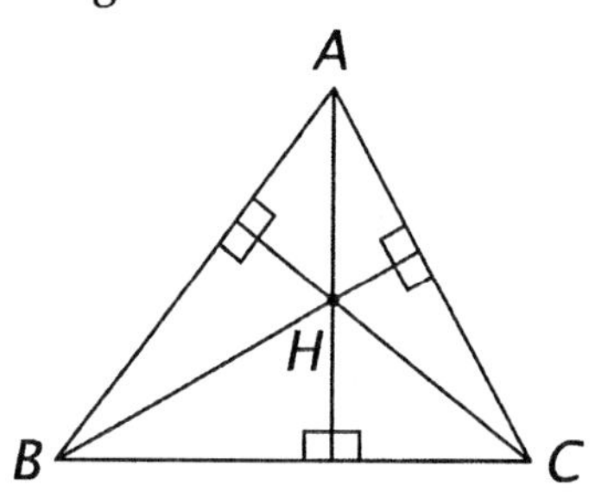

It appears below with three words omitted.

> Although the _______ are three,
> Remarks my daughter _______,
> One point'll lie on all of them:
> The _______ H'll.

Two of the missing words are geometric and one is his daughter's name. What do you think the three words are?

TURN OVER

6. This figure reveals that quadrilateral ABCD is cyclic.

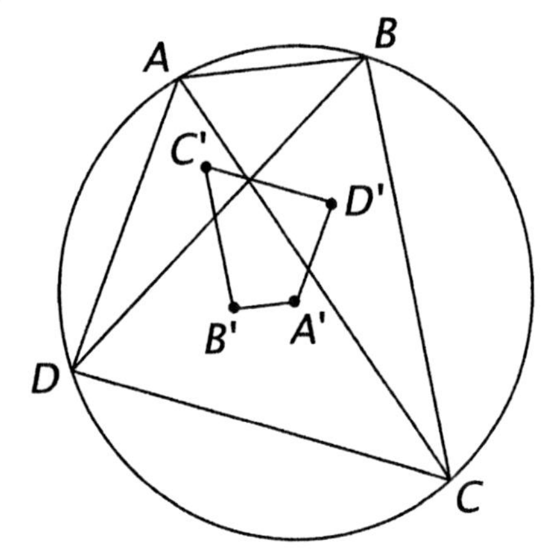

a) What does the word *cyclic* mean?
b) What can you conclude about the opposite angles of ABCD?

Point D′ is the centroid of ΔABC, A′ is the centroid of ΔBCD, B′ is the centroid of ΔACD, and C′ is the centroid of ΔABD.

c) What is a centroid of a triangle?

Quadrilateral A′B′C′D′ is related to quadrilateral ABCD in a special way.

d) What do you think it is?

7. In this figure, equilateral ΔBCD and its circumcircle have been constructed on side BC of ΔABC; line AD intersects the circle at P.

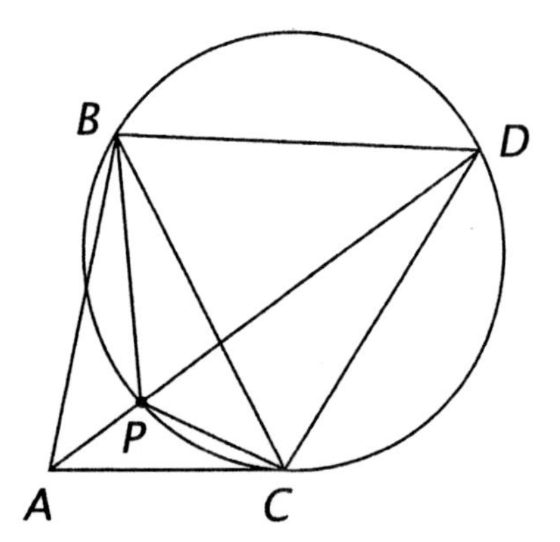

What can you conclude about the measures of the following angles? Explain each answer.

a) ∠BPD.
b) ∠CPD.
c) ∠BPA.
d) ∠CPA.
e) ∠BPC.

Point P is called the Steiner Point of ΔABC.

f) What do these results reveal about the Steiner Point of a triangle?

8. In this figure, the cevians AD, BE, and CF are perpendicular to the sides of ΔABC.

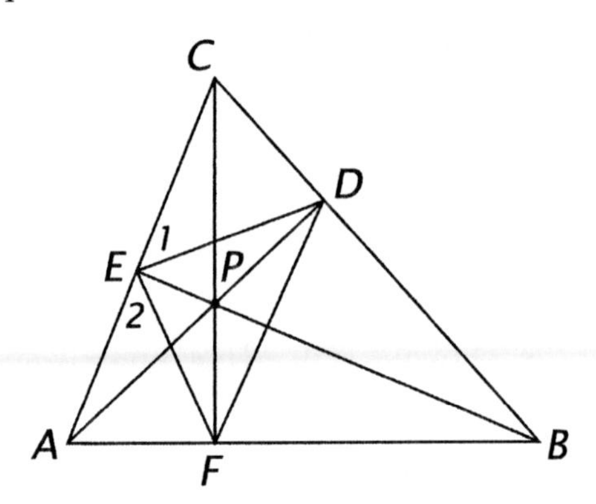

a) State the theorem that tells us that they are concurrent.

Point P is also the incenter of ΔDEF.

b) What are cevians DA, EB, and FC called with respect to ΔDEF?
c) What can you conclude about ∠1 and ∠2? Explain.
d) If a billiard table were built in the shape of ΔABC, where do you think a ball at D would go if it were hit in the direction DE?

GEOMETRY: Test on Chapter 14

Name______________________________

Regular Polygon Formulas

The perimeter of a regular polygon having n sides is $2Nr$, in which $N = n \sin \frac{180}{n}$ and r is its radius.

The area of a regular polygon having n sides is Mr^2, in which $M = n \sin \frac{180}{n} \cos \frac{180}{n}$ and r is its radius.

1. Read the following statements carefully and mark them true or false.

   a) The circumference of a circle is the limit of the perimeters of the inscribed regular polygons.
   b) Every polygon that is equilateral is also cyclic.
   c) An apothem of a regular polygon is perpendicular to one of its sides.
   d) If two arcs have equal measures, they must also have equal lengths.
   e) The ratio of the areas of two circles is equal to the ratio of their radii.

2. This figure shows a regular nonagon inscribed in a circle.

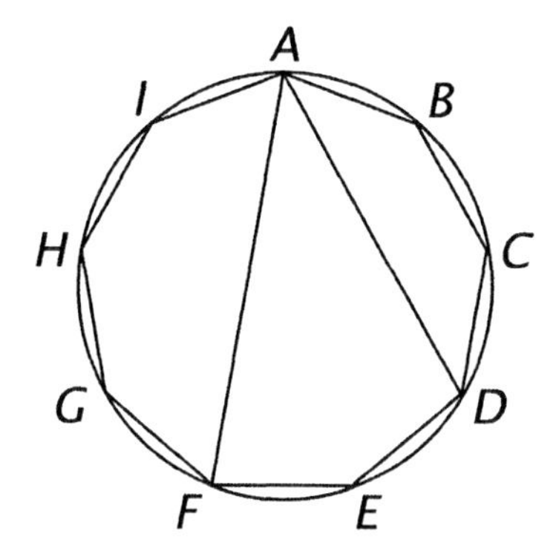

   Find the measures of the following arcs and angles.

   a) $m\overset{\frown}{AB}$.
   b) $m\overset{\frown}{BD}$.
   c) $\angle BAD$.
   d) $\angle B$.

   What can you conclude about

   e) BC and AD? Explain.
   f) ABCD? Explain
   g) ADEF? Explain.

3. A goat is tied with a chain to one corner of a rectangular field, 308 feet long and 204 feet wide. The area that the goat can graze is one-half that of the field.

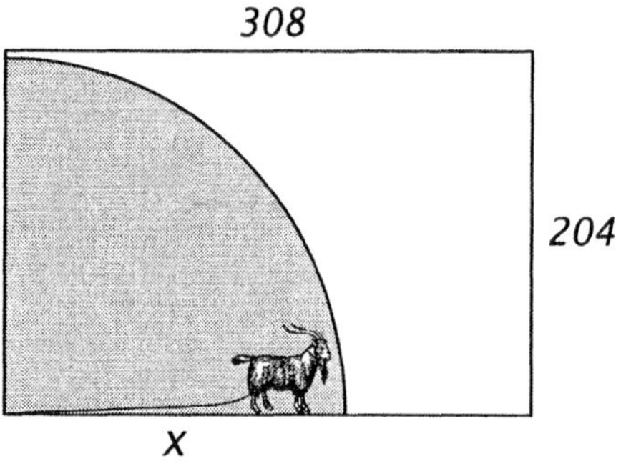

   a) How long is the chain?
   b) What is the length of the arc along the edge of the goat's grazing area?

4. This design, from a book published in 1588, was once thought to have mystical powers.

   Given that the diameter of the large circle is 4 units, find the total area of the 16 white regions inside it. Leave your answer in terms of π.

5. In his book *Measurement of a Circle*, Archimedes compared the areas of the circle and square in this figure.

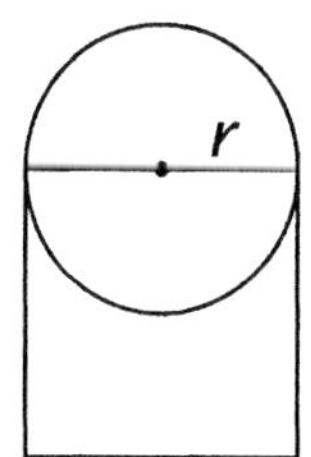

   He said that $\frac{\alpha \text{ circle}}{\alpha \text{ square}} = \frac{11}{14}$.

   If this were correct, to what number would π be equal?

TURN OVER

6. The Castel del Monte is shown on this postal stamp.

Built in Italy in the 13th century, its basic floor plan is shown in the figure below. Its inner and outer walls lie along the sides of two regular octagons.

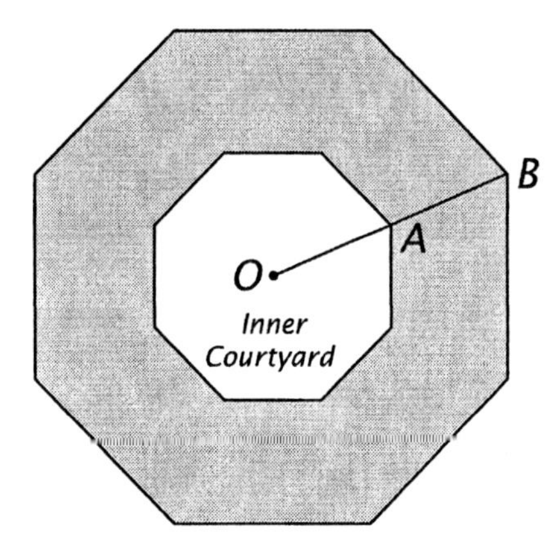

Given that the radius of the inner wall, OA, is 35 feet and that the radius of the outer wall, OB, is 70 feet, find each of the following.

a) The area of the inner courtyard (the smaller octagon).
b) The perimeter of the outer wall.
c) The area enclosed between the outer and inner walls to the nearest hundred square feet.

7. This figure shows a regular dodecagon divided into 12 congruent triangles.

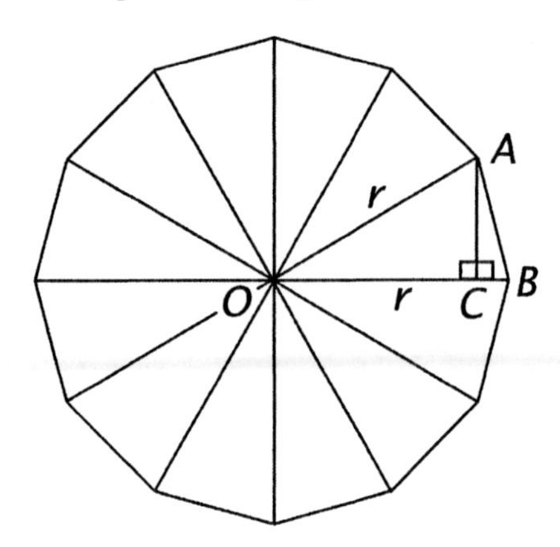

Notice that in one of these triangles, ΔAOB, $OA = OB = r$, the radius of the dodecagon; AC has been drawn perpendicular to OB.

a) Find $\angle AOB$.
b) Find $\angle OAC$.
c) Find AC in terms of r.
d) Find $\alpha\Delta AOB$ in terms of r.
e) Find the area of the dodecagon in terms of r.

f) Why is the formula for the area of a regular dodecagon so much like the formula for the area of a circle?

GEOMETRY: Test on Chapter 15

Name_______________________________

1. Read the following statements carefully and mark them true or false.

   a) If two planes are perpendicular to a third plane, the planes are parallel to each other.
   b) Cavalieri's Principle is a way to prove that two solids have the same volume.
   c) The lateral faces of every prism are parallelograms.
   d) The cross sections of a sphere are circles.
   e) All cones are similar.

2. During World War 2, the British government considered using icebergs as unsinkable aircraft carriers.

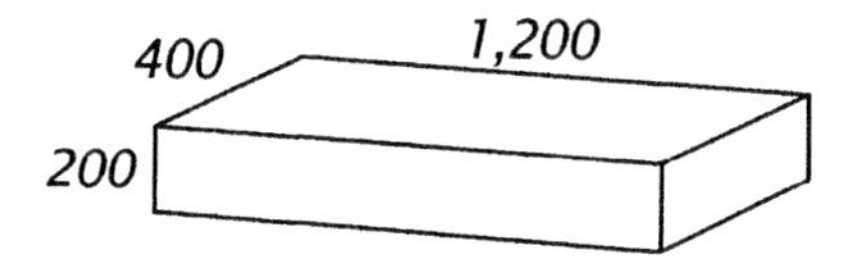

   Suppose that an iceberg has the shape of a rectangular solid 400 m wide, 1,200 m long and 200 m thick.

   a) Find the area of its top face.
   b) Find its volume.
   c) Find its total surface area.

3. The figures below show side views of three vases.

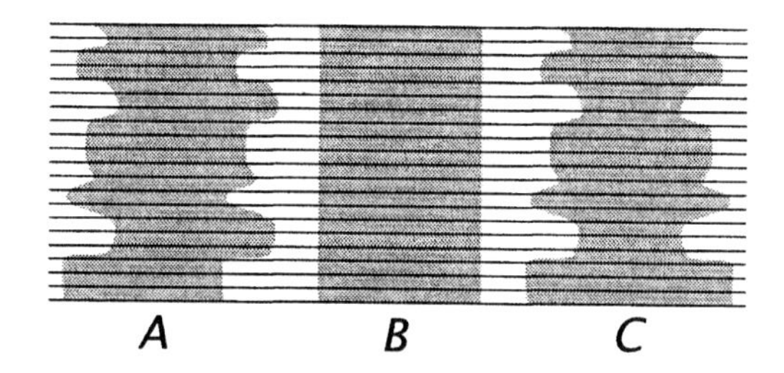

   The vase in the middle has the shape of a right cylinder.

   a) Write an expression for the volume of the cylinder in terms of the radius of its base, r, and its altitude, h.
   b) Which of the other two vases looks as if it might have the same volume? Explain.

4. The following is a sample problem from a test given to French students when they have finished the 9th grade.

   The cone shown below has a height of 9 cm. Its base has a radius of 3 cm.

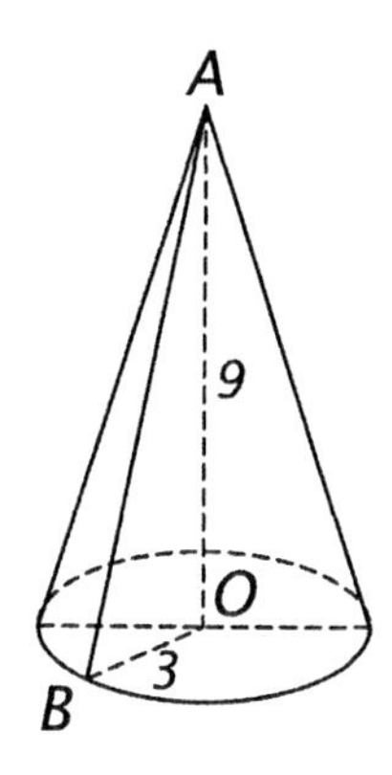

   a) Find the exact length of AB.
   b) Find the measure of $\angle$BAO to the nearest degree.
   c) Find the volume of the cone to the nearest cm^3.

5. Obtuse Ollie blew up a spherical balloon until it popped. At its biggest, the balloon had a surface area of 2,463 square inches.

   a) What was its radius just before it popped?
   b) What was its volume?

TURN OVER

6. Pictured on the back of a $1 bill is the frustum of a square pyramid, the part of the pyramid included between its base and a plane parallel to its base.

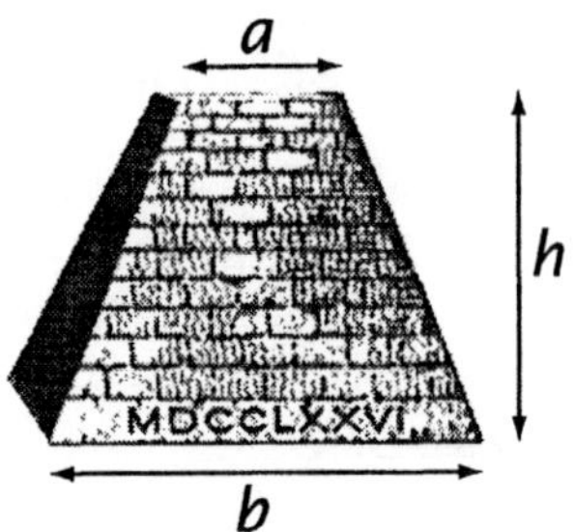

The volume of the frustum is

$$\frac{1}{3}(a^2 + ab + b^2)h$$

where its dimensions are a, b, and h as shown on the figure.

a) What does b^2 represent?
b) What does a^2 represent?
c) What does the volume expression become if $a = 0$?
d) For what type of geometric solid does your answer to part c give the volume?
e) What does the volume expression become if $a = b$?
f) For what type of geometric solid does your answer to part e give the volume?

7. To film a horror movie, you have been asked to make two models of a monster. The models are to be similar, with heights of 16 feet and 40 feet.

a) How many times the height of the smaller monster is the height of the larger one?
b) How many times the surface area of the smaller monster will the surface area of the larger one be?
c) How many times the volume of the smaller monster will the volume of the larger one be?

You have already made the smaller monster. It is covered with 60 square feet of fur, weighs 400 pounds, and has arms that are 10 feet long.

d) How much fur do you need for the larger monster?
e) How much would you expect the larger monster to weigh?
f) How long should its arms be?

GEOMETRY: Test on Chapter 16

Name________________________________

Comparison of the three geometries

Statement	Euclid	Lobachevsky	Riemann
Through a point not on a line, there is	exactly one parallel to the line.	more than one parallel to the line.	no parallel to the line.
The summit angles of a Saccheri quadrilateral are	right.	acute.	obtuse.
The sum of the angles of a triangle is	180°.	less than 180°.	more than 180°.

Theorems true in Lobachevskian geometry

The summit of a Saccheri quadrilateral is longer than its base.
A midsegment of a triangle is less than half as long as the third side.
The sum of the angles of a quadrilateral is less than 360°.
If two triangles are similar, they must also be congruent.

1. Read the following statements carefully and mark them true or false.

   a) In Euclidean geometry, if one of the summit angles of a birectangular quadrilateral is acute, the other summit angle must be obtuse.
   b) Each angle of an equilateral triangle in Lobachevskian geometry is less than 60°.
   c) Einstein's theory of relativity implies that physical space is non-Euclidean in nature.
   d) In Riemannian geometry, the summit of a Saccheri quadrilateral is shorter than its base.
   e) In sphere geometry, through a point not on a line, there is exactly one line perpendicular to the line.

2. This beach ball has been divided by great circles into congruent isosceles triangles.

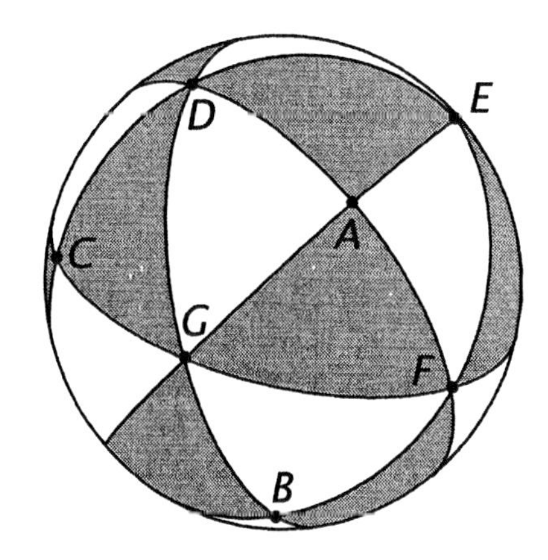

The angles at A, B, and C are right angles and the angles at D, E, F, and G are 60°.

   a) What relation do the acute angles of a right triangle have in Euclidean geometry?
   b) Does the same relation hold in sphere geometry? Explain.
   c) If the diagonals of a quadrilateral in Euclidean geometry bisect each other, what can you conclude?
   d) Does the same conclusion hold in sphere geometry? Explain.

3. In this figure, ΔABC represents a triangle in Lobachevskian geometry and $\angle 1$ is one of its exterior angles.

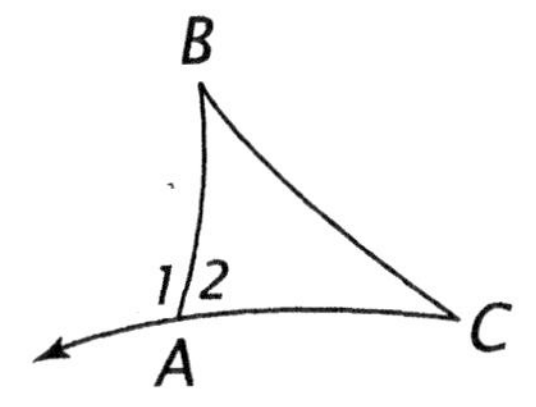

Copy and complete the following statements.

   a) $\angle 2 + \angle B + \angle C$? 180°.
   b) $\angle 1 + \angle 2$? 180°.
   c) Use your answers to a and b to derive a relation between $\angle B + \angle C$ and $\angle 1$.
   d) State a theorem in Lobachevskian geometry suggested by your answer to c.

TURN OVER

4. This figure represents a plan for the floor of a tennis court in Lobachevskian geometry.

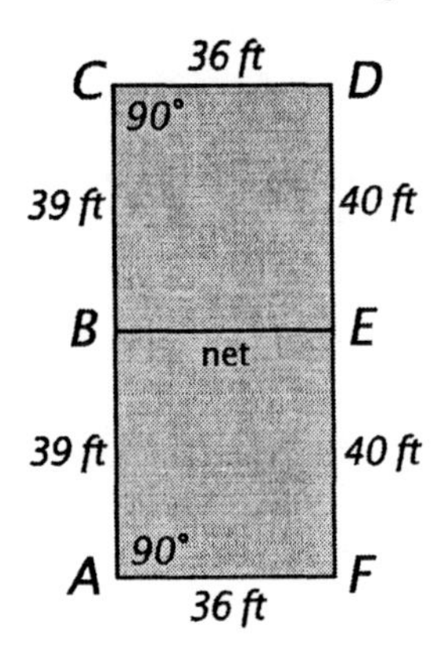

a) What kind of quadrilateral is ACDF? (Remember that this is in Lobachevskian geometry.)
b) Why is the net perpendicular to the sides of the court?
c) What kind of quadrilaterals are ABEF and BCDE?
d) What kind of angles are ∠F and ∠D? Explain.
e) What can you conclude about the length of the net, BE? Explain.

5. A theorem that is true in both Euclidean and non-Euclidean geometries is:

If the legs of a birectangular quadrilateral are unequal, the summit angles opposite them are unequal in the same order.

Complete the following proof of this theorem by giving a reason for each of the lettered statements.

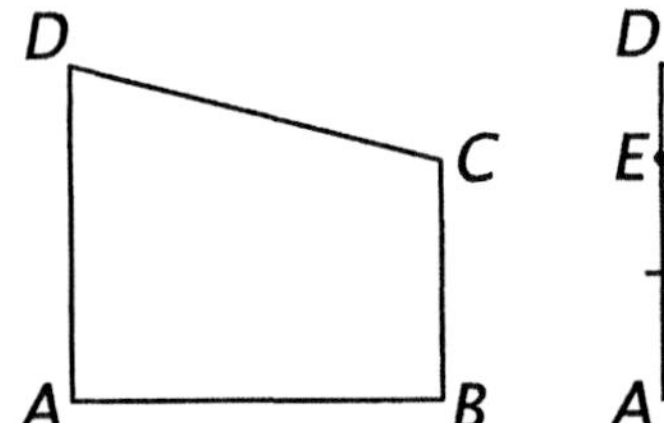

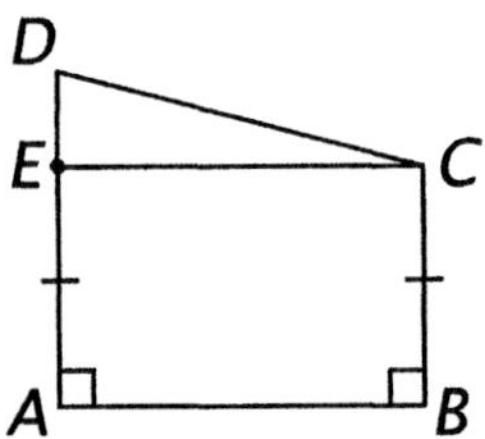

Given: Birectangular quadrilateral ABCD with base AB; AD > BC.
Prove: ∠C > ∠D.

Proof.
ABCD is a birectangular quadrilateral with base AB; AD > BC. AD ⊥ AB and BC ⊥ AB because the legs of a birectangular quadrilateral are perpendicular to its base.

a) Choose point E on AD so that AE = BC.

Draw EC.

b) ABCE is a Saccheri quadrilateral.
c) ∠AEC = ∠ECB.
d) ∠AEC > ∠D.
e) ∠ECB > ∠D.
f) ∠DCB = ∠DCE + ∠ECB.
g) ∠DCB > ∠ECB.
h) ∠DCB > ∠D (or, ∠C > ∠D).

6. In this figure, the vertices of ΔABC are the midpoints of the sides of ΔDEF.

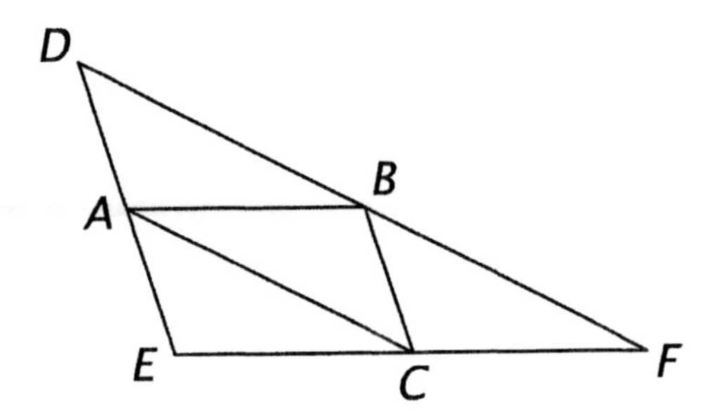

How does the perimeter of ΔABC compare to the perimeter of ΔDEF

a) in Euclidean geometry? Explain.
b) in Lobachevskian geometry?
c) What can you conclude about the directions of the sides of the two triangles in Euclidean geometry?
d) Does the same conclusion apply in Riemannian geometry? Explain

In Euclidean geometry, ΔABC is similar to ΔDEF.

e) Can this also be true in Lobachevskian geometry? Explain.

GEOMETRY: Midyear Exam—Part 1

Do as indicated.

1. Simplify: $x + x + x$.
2. Use the distributive rule to eliminate the parentheses: $x^2(4 - x)$.
3. Solve for x: $3(x + 7) = -12$.
4. Subtract: $(9x + y) - (x - y)$.
5. Factor: $5x^2 + 14x - 3$.
6. Reduce to lowest terms: $\frac{x-3}{x^2-9}$.
7. Write as a single fraction in lowest terms: $\frac{x+2}{4} - \frac{x}{8}$.
8. Write $\sqrt{72}$ in simple radical form.

Formulas are important in geometry as well as in algebra. Explain what each of these formulas means.

9. $c = \pi d$.
10. $A = s^2$.

The following questions refer to this statement:

All porcupines have long quills.

11. Write the statement in "if-then" form.
12. Does it follow that an animal that has long quills is a porcupine?
13. What relation does this idea have to the original statement?

Write the letter of the correct answer on your answer sheet.

14. If two angles of one triangle are equal to two angles of another triangle,
    a) the triangles must be congruent.
    b) the sides opposite them are equal.
    c) the triangles are equiangular.
    d) the third pair of angles must be equal.
15. If $a > b$, which one of the following inequalities must be true?
    a) $a^2 > b^2$.
    b) $a + c > b + c$.
    c) $ac > bc$.
    d) $a + b > 0$.
16. If a quadrilateral is equiangular,
    a) it is also equilateral.
    b) it is a square.
    c) it is a rectangle.
    d) it is concave.
17. Two angles are a linear pair. Which of the following must be true?
    a) The angles are equal.
    b) The angles are right angles.
    c) The angles are supplementary.
    d) All of these must be true.
18. If the legs of one right triangle are equal to the legs of another right triangle, which one of the following could be used to prove the triangles congruent?
    a) SAS.
    b) ASA.
    c) SSS.
    d) HL.
19. Which of the following could be the lengths of the sides of a triangle?
    a) 3, 7, 11.
    b) 4, 4, 8.
    c) 5, 12, 13.
    d) 6, 9, 15.
20. Two lines are parallel if the interior angles that they form on the same side of a transversal
    a) are right angles.
    b) are equal.
    c) are complementary.
    d) form a linear pair.
21. If all of the exterior angles of a triangle are obtuse, the triangle must be
    a) obtuse.
    b) equiangular.
    c) acute.
    d) scalene.
22. Which one of the following statements about the diagonals of a parallelogram is always true?
    a) They are equal.
    b) They are longer than the sides.
    c) They are perpendicular.
    d) They bisect each other.
23. If the sides of a triangle have lengths x, x, and y, its perimeter is
    a) $x^2 + y$.
    b) $2xy$.
    c) x^2y.
    d) $2x + y$.

TURN OVER

Read the following statements carefully. If a statement is always true, write *true*. If not, *do not write false*. Instead, write a word or words that could replace the underlined word to make the statement true. Some of the questions in this section may have more than one correct answer; do not make a change in any statement that is always true however.

24. An obtuse triangle has two obtuse angles.
25. Three noncollinear points determine a plane.
26. The base angles of an isosceles trapezoid are equal.
27. The converse of every theorem is true.
28. The points on a line can be numbered so that positive number sums measure distances.
29. The diagonals of a rhombus are equal.
30. The word line is an undefined term in geometry.
31. If the x-coordinate of a point is 0, the point is on the x-axis.
32. A pentagon is a polygon that has eight sides.
33. Two angles are vertical angles if the sides of one angle are opposite rays to the sides of the other.
34. If $\angle A$ and $\angle B$ are supplementary and $\angle B$ and $\angle C$ are supplementary, than $\angle A$ and $\angle C$ are supplementary.
35. According to the Betweenness of Points Theorem, if A-B-C, then AB = BC.
36. To prove a theorem indirectly, we begin by assuming that the opposite of its conclusion is true.
37. If a triangle has two equal sides, it must be equilateral.
38. Two lines that do not intersect must be parallel.
39. The supplement of an acute angle is acute.
40. Through a point not on a line, there is exactly one line parallel to the given line.
41. The "Whole Greater than Part" Theorem says that if $a > 0$, $b > 0$, and $a + b = c$, then $a > c$ and $b > c$.
42. In a right triangle, the hypotenuse must be the longest side.
43. A square has two lines of symmetry.

Use your straightedge and compass to make the following constructions on your answer sheet.

44. Bisect $\angle A$.
45. Through D, construct a line perpendicular to BC.

GEOMETRY: *Answer Sheet for* Midyear Exam–Part 1

Name______________________________

1. ____________ 5. ____________

2. ____________ 6. ____________

3. ____________ 7. ____________

4. ____________ 8. ____________

9. ______________________________

____

10. ______________________________

____

11. ______________________________

____

12. ______________________________

13. ______________________________

14. ______ 18. ______ 22. ______

15. ______ 19. ______ 23. ______

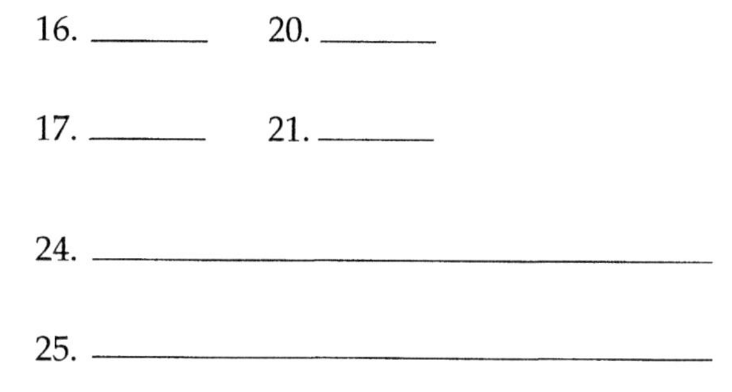

16. ______ 20. ______

17. ______ 21. ______

24. ______________________________

25. ______________________________

26. ______________________________

27. ______________________________

28. ______________________________

29. ______________________________

30. ______________________________

31. ______________________________

32. ______________________________

33. ______________________________

34. ______________________________

35. ______________________________

36. ______________________________

37. ______________________________

38. ______________________________

39. ______________________________

40. ______________________________

41. ______________________________

42. ______________________________

43. ______________________________

44/45.

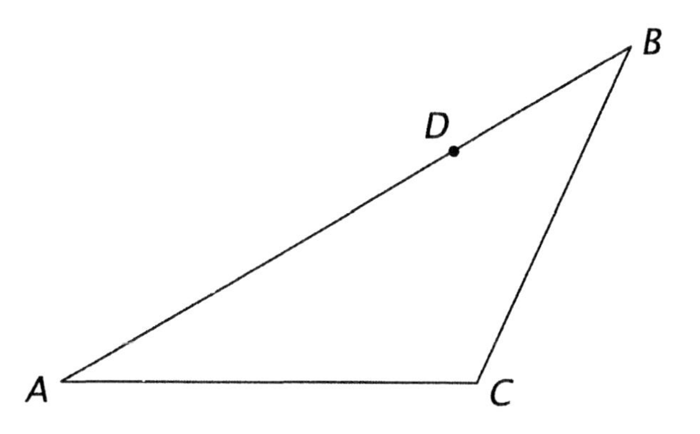

GEOMETRY: Midyear Exam—Part 2

Two words that appear near each other in the dictionary are "hypotenuse" and "hypothesis."

1. To what does the word "hypotenuse" refer?
2. To what does the word "hypothesis" refer?

Here is a familiar figure.

3. Except for his tail, what kind of symmetry does the figure have?
4. What is a simple test for this kind of symmetry?

This figure shows a protractor centered on O, the endpoint of four rays. The coordinates of OA and OC are shown.

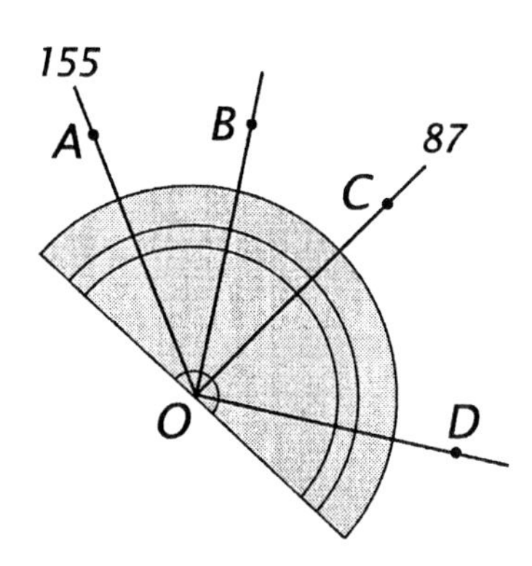

5. Find $\angle$AOC.
6. Find $\angle$BOC, given that OB bisects $\angle$AOC.
7. Find the coordinate of OB.
8. Find $\angle$COD, given that $\angle$BOC and $\angle$COD are complementary.
9. Find the coordinate of OD.

In this figure, BD > AD and A-D-C. Write the appropriate symbol (>, =, or <) to make each of the following statements true.

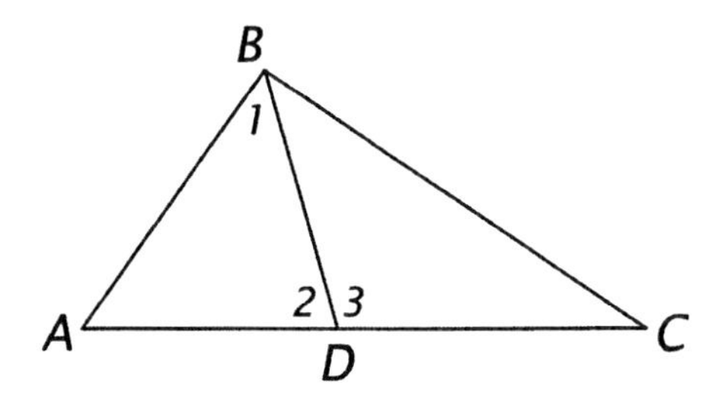

10. $\angle 1$? $\angle$A.
11. $\angle 3$? $\angle$A + $\angle 1$.
12. AC ? AD + DC.
13. BC ? BD + DC.
14. $\angle 2$? $\angle$C.

In ΔABC, AD = DC = BE = EC.

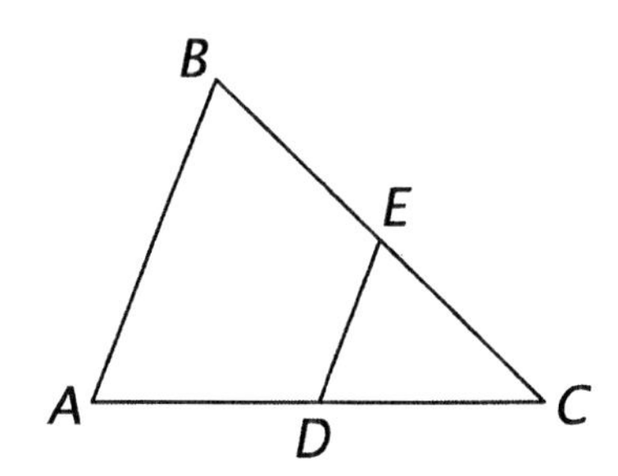

15. What can you conclude about ΔABC?
16. Why is DE $\parallel$ AB?
17. Why is $\angle$A = $\angle$EDC?
18. What can you conclude about ABED?
19. Why is $\angle$ADE = $\angle$DEB?

TURN OVER

This figure from an old book shows a method for measuring the height of a tower.

Refer to the information marked on the version of it below to answer the following questions.

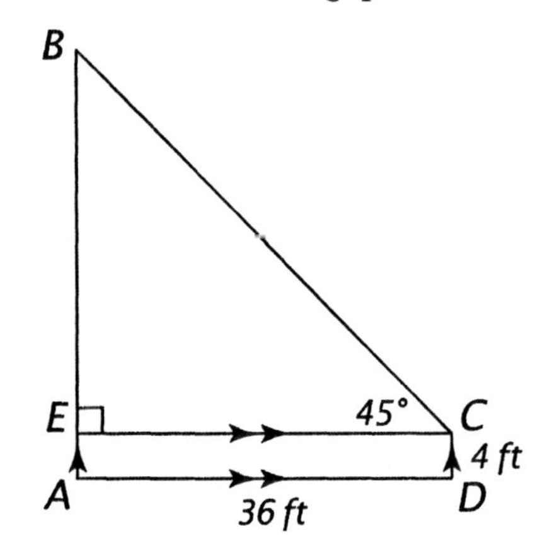

20. What can you conclude about ADCE?
21. What can you conclude about ΔBCE?
22. Why is EC = AD?
23. What is the height of the tower?

This figure contains four congruent quadrilaterals. The equal angles are indicated by the arcs.

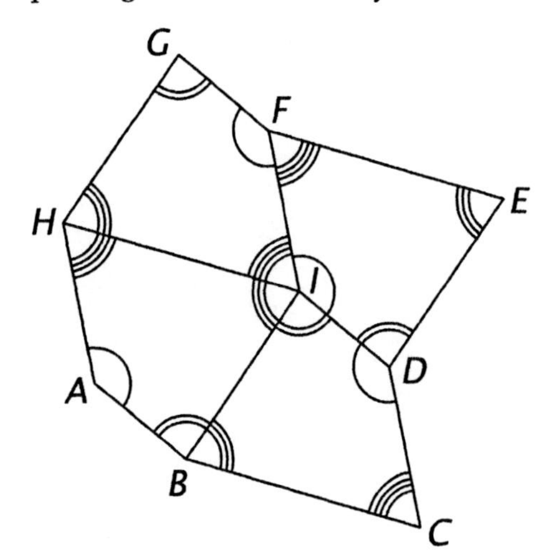

Can you conclude from this information that

24. any of the segments in the figure are equal? Explain.
25. any of the segments in the figure are parallel? Explain.
26. What theorem is suggested by the four angles surrounding point I?

This figure shows an ordinary file card that has been folded along MC and MD and opened flat again. As a result, ΔACM ≅ ΔPCM and ΔBDM ≅ ΔPDM.

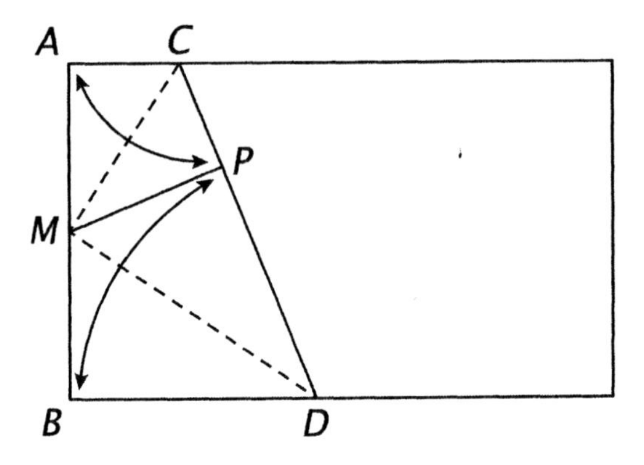

Can you conclude from this information that

27. M is the midpoint of AB? Explain.
28. C, P, and D are collinear? Explain.
29. CM ⊥ MD? Explain.

Write a complete proof for the following.

30.

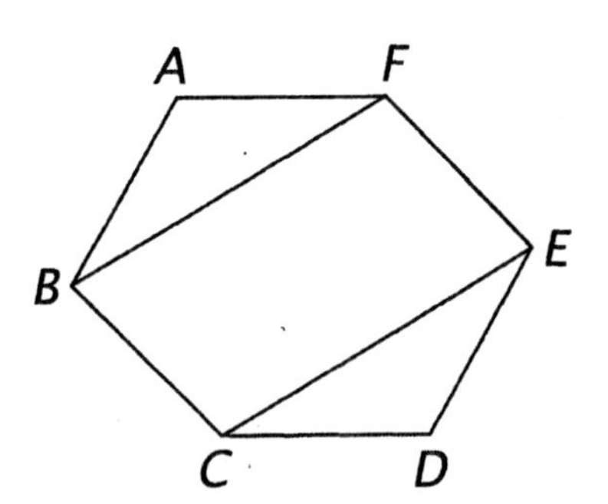

Given: AB = BC = CD = DE = EF = FA and ∠A = ∠D.

Prove: BC || FE.

GEOMETRY: *Answer Sheet for* Midyear Exam–Part 2

Name ______________________________

1. ______________________________
   ____
2. ______________________________
   ____
3. ______________________________
4. ______________________________
   ____

5. ______ 9. ______ 13. ______

6. ______ 10. ______ 14. ______

7. ______ 11. ______

8. ______ 12. ______

15. ______________________________
16. ______________________________
    ____
17. ______________________________
    ____
18. ______________________________
19. ______________________________
    ____
20. ______________________________
21. ______________________________
22. ______________________________
    ____
23. ______________________________
24. ______________________________
    ____
25. ______________________________
    ____
    ____
26. ______________________________
    ____
27. ______________________________
    ____
    ____
28. ______________________________
    ____
    ____
29. ______________________________
    ____
    ____
30.

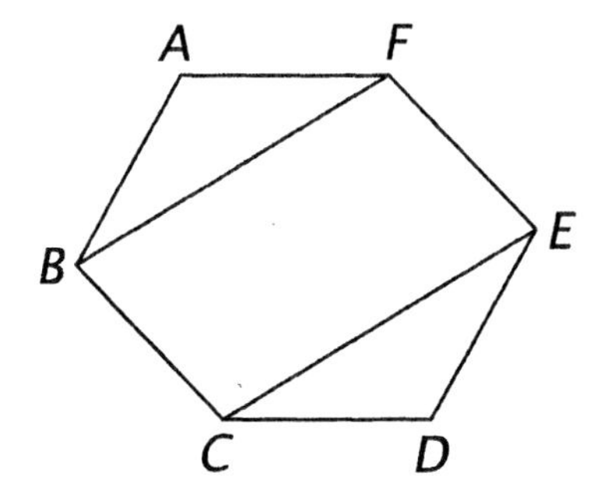

Given: AB = BC = CD = DE = EF = FA and $\angle A = \angle D$.
Prove: $BC \parallel FE$.

GEOMETRY: Final Exam—Part 1

Read the following statements carefully. If a statement is always true, write *true*. If not, *do not write false*. Instead, write a word or words that could replace the underlined word or words to make the statement true. Some of the questions in this section may have more than one correct answer; do not make a change in any statement that is always true however.

1. Every triangle has an incircle.
2. Cavalieri's Principle is used to prove that two geometric solids have equal areas.
3. The secant segments to a circle from an external point are equal.
4. The ratio of the perimeters of two similar polygons is equal to the square of the ratio of the corresponding sides.
5. A radius of a regular polygon is a perpendicular line segment from its center to one of its sides.
6. Every cross section of a cylinder is a circle.
7. The circumference of a circle is the limit of the perimeters of the inscribed regular polygons.
8. A cevian of a triangle is a line segment that connects a vertex to a point on the opposite side.
9. If b is the geometric mean between a and c, then $b^2 = ac$.
10. The area of a parallelogram is the product of its length and width.
11. The ratio of the circumference to the radius of a circle is π.
12. In a 30°-60° right triangle, the hypotenuse is twice the longer leg.
13. Corresponding angles of similar triangles have the same ratio as the corresponding sides.
14. The perpendicular bisectors of the chords of a circle (in the plane of the circle) are collinear.
15. The sine of an acute angle of a right triangle is the ratio of the length of the opposite leg to the length of the adjacent leg.
16. If e is the length of one of the edges of a cube, the volume of the cube is $6e^2$.

The following problem appears on a Babylonian clay tablet from about 1600 B.C.:

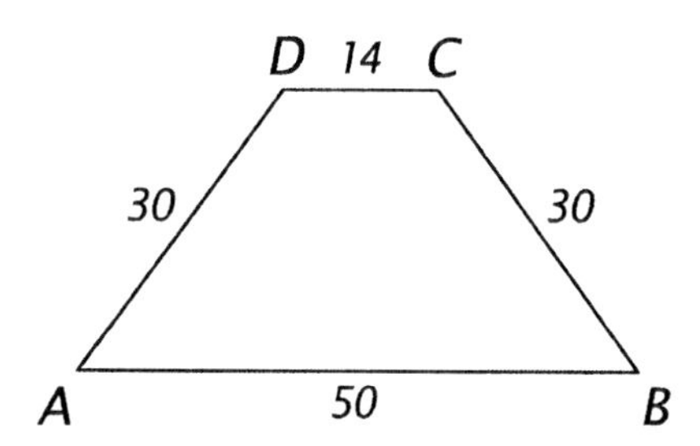

Find the area of an isosceles trapezoid whose bases are 14 and 50 and whose legs are 30.

A way to solve the problem is suggested by the figure below.

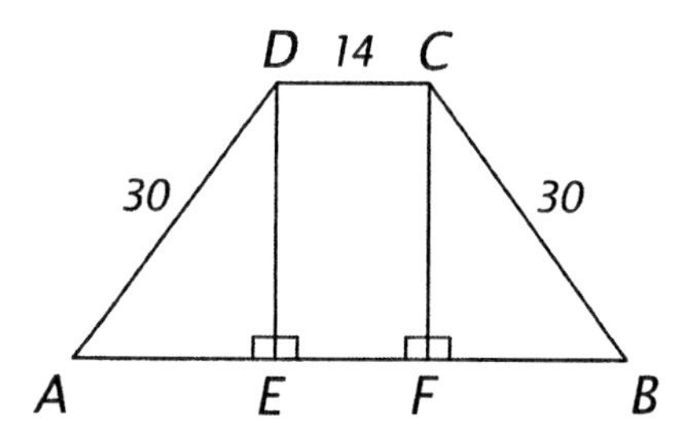

17. Why is $\angle A = \angle B$?
18. Why is $\Delta AED \cong \Delta BFC$?

Find each of the following.

19. AE.
20. DE.
21. αABCD.

TURN OVER

This figure shows a way to measure the height of a cliff.

22. How does the length of the radius of the circle drawn around the post appear to have been chosen?
23. What kind of triangles are being used?
24. What is assumed about the rays of sunlight that cast the shadows?
25. What relation do the triangles have to each other?
26. How is the height of the cliff found?

One of our theorems about circles is stated in Euclid's *Elements* in the following way:

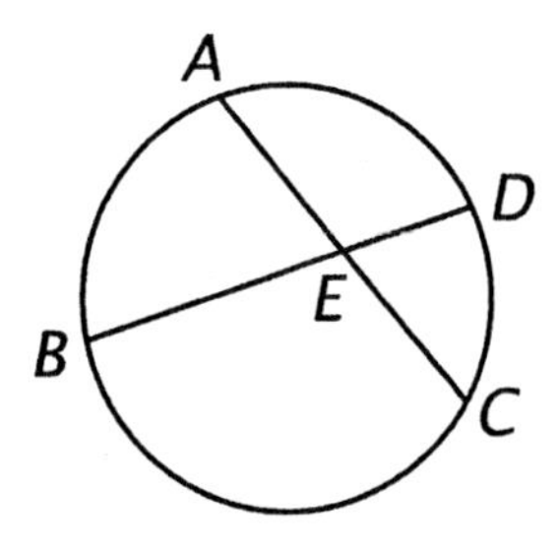

"If in a circle two straight lines cut one another, the rectangle contained by the segments of the one is equal to the rectangle contained by the segments of the other."

Our name for this theorem is based on the part that says "two straight lines cut one another."

27. What is it?

There are no rectangles in Euclid's figure for the theorem.

28. What did he mean in saying that the rectangles contained by the segments are equal?
29. What would the "rectangles" become if the two chords were diameters of the circle?

This figure shows the way in which NASA uses aircraft to map ground by means of radar.

AB = 7.3 km, ∠ABC = 23°, and ∠CBD = 39°.

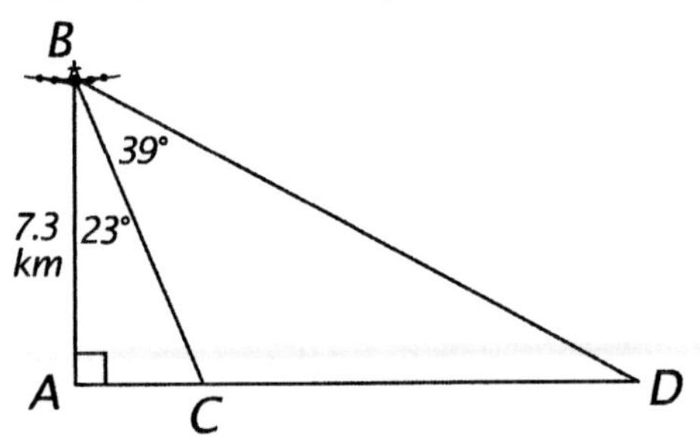

Find each of the following distances to the nearest 0.1 km.

30. AC.
31. AD.
32. CD.

These figures illustrate a construction that is impossible to do for most angles.

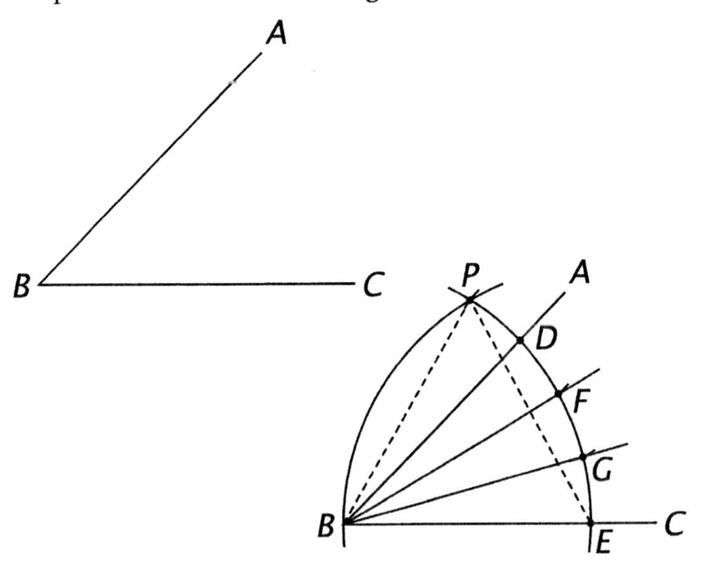

The first figure shows a 45° angle, ∠ABC. The second figure shows the completed construction.

An arc centered at B has been drawn to intersect the sides of ∠ABC at D and E. A second arc centered at E and having radius EB has been drawn to intersect the first arc at P.

33. What can you conclude about ∠PBE? Explain.
34. What can you conclude about ∠PBD and $m\overset{\frown}{PD}$?

Points F and G were marked on $\overset{\frown}{DE}$ so that $m\overset{\frown}{PD} = m\overset{\frown}{DF} = m\overset{\frown}{FG}$.

35. What can you conclude about BF and BG with respect to ∠ABC?

CONTINUED ON NEXT PAGE

Make the following construction as accurately as you can *on the figure on your answer sheet.*

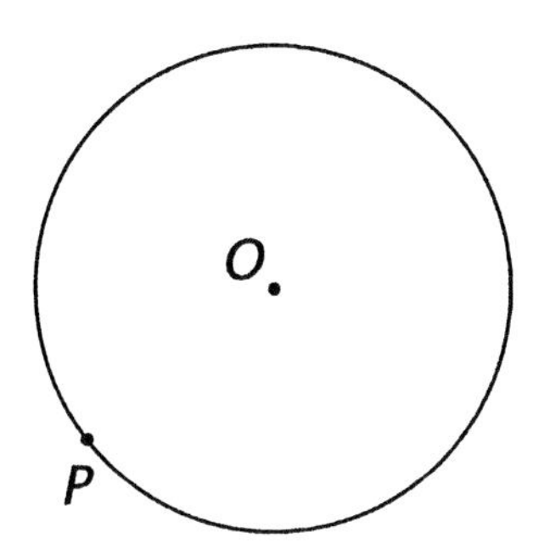

36. Construct a line through point P that is tangent to circle O.

Water is drawn from a well by means of a rope wound around a drum.

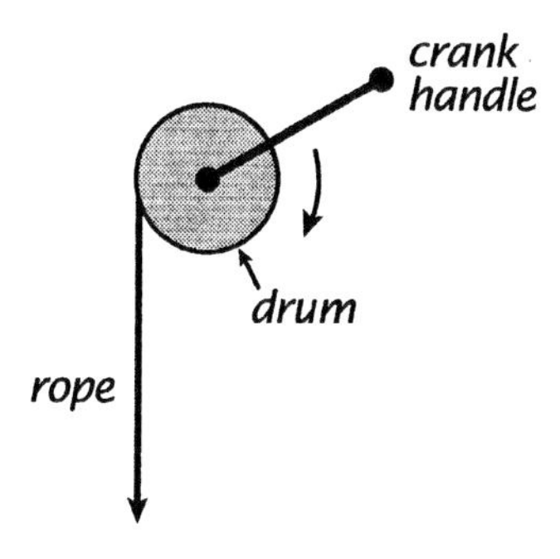

37. If the diameter of the drum is 6 inches, how far would a bucket of water rise when the drum is turned through one revolution?

38. Approximately how many turns would it take to raise a bucket of water from the bottom to the top of a well 25 feet deep?

Concrete is measured in cubic yards. (1 yard = 3 feet.)

39. How many cubic feet are in one cubic yard?

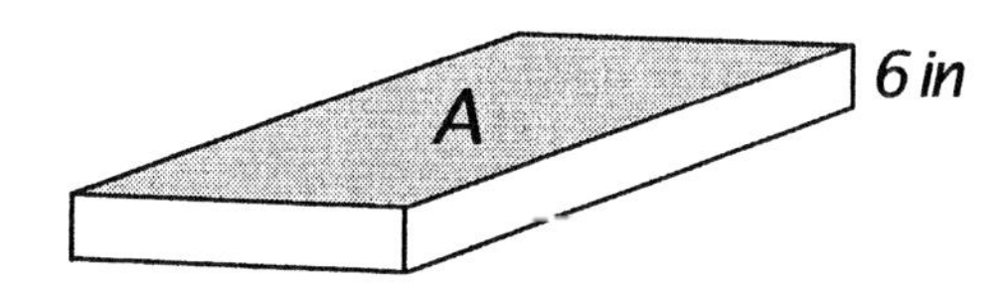

One cubic yard of concrete is used to make a slab in the shape of a rectangular solid.

40. How many square feet will the slab cover if it is 6 *inches* thick?

GEOMETRY: *Answer Sheet for* Final Exam–Part 1

Name________________________________

1. ______________________
2. ______________________
3. ______________________
4. ______________________
5. ______________________
6. ______________________
7. ______________________
8. ______________________
9. ______________________
10. ______________________
11. ______________________
12. ______________________
13. ______________________
14. ______________________
15. ______________________
16. ______________________
17. ______________________
    ____
18. ______________________
19. ______________________
20. ______________________
21. ______________________
22. ______________________
    ____
23. ______________________
24. ______________________
25. ______________________
26. ______________________
    ____
27. ______________________
28. ______________________
    ____
    ____
29. ______________________
30. ______________________
31. ______________________
32. ______________________
33. ______________________
    ____
34. ______________________
35. ______________________

TURN OVER

36.

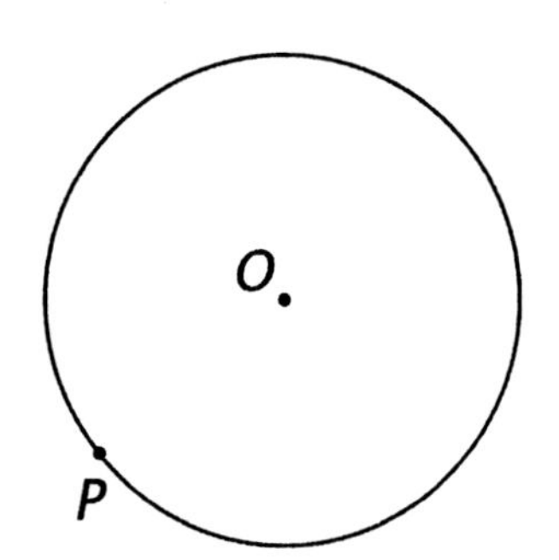

37. ______________________________

38. ______________________________

39. ______________________________

40. ______________________________

GEOMETRY: Final Exam—Part 2

Write the letter of the correct answer on your answer sheet.

1. Which of the following sets of numbers can be the lengths of the sides of a right triangle?
   a) 4, 8, 9.
   b) 9, 16, 25.
   c) 20, 21, 29.
   d) 12, 12, 17.

2. Two triangles that have equal bases and equal altitudes must
   a) have equal sides.
   b) have equal areas.
   c) be congruent.
   d) have equal perimeters.

3. Which one of the following is *false*?
   If an altitude is drawn to the hypotenuse of a right triangle,
   a) it is perpendicular to the hypotenuse.
   b) it must lie inside the triangle.
   c) it forms two triangles similar to the original triangle.
   d) it is the geometric mean between the legs of the triangle.

4. Which of the following polygons are similar?
   a) All isosceles triangles.
   b) All squares.
   c) All rectangles.
   d) All pentagons.

5. Ceva's Theorem is used to
   a) find the area of a triangle given the lengths of its sides.
   b) find the balancing point of a triangle.
   c) prove that a triangle is a right triangle.
   d) prove that lines in a triangle are concurrent.

6. A central angle of a regular decagon has a measure of
   a) 18°.
   b) 30°.
   c) 36°.
   d) 72°.

7. Which one of the following statements about a regular pyramid is *not necessarily true*?
   a) Its base is a regular polygon.
   b) Its lateral edges are equal.
   c) Its lateral faces are equilateral triangles.
   d) Its altitude connects its vertex to the center of its base.

8. Which one of the following polygons is not always cyclic?
   a) A scalene triangle.
   b) A rectangle.
   c) A regular hexagon.
   d) A parallelogram.

What is each of the following formulas used for?

9. $d = s\sqrt{2}$.

10. $A = \frac{\sqrt{3}}{4}s^2$.

11. $p = 2Nr$, in which $N = n \sin\frac{180}{n}$.

12. $V = \frac{4}{3}\pi r^3$.

The hypotenuse of this right triangle is equal to 1.

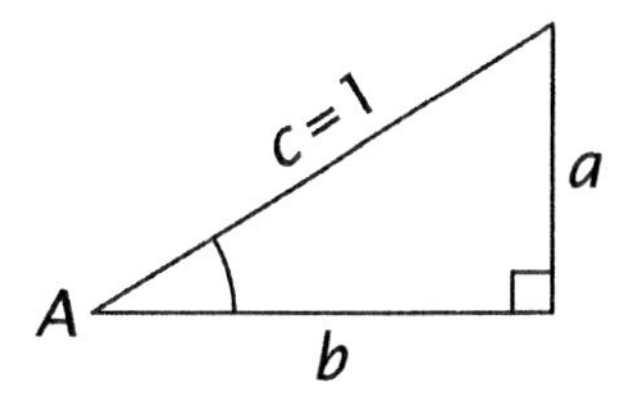

Which side of the triangle is equal to

13. sin A?

14. cos A?

15. What can you conclude about $(\sin A)^2 + (\cos A)^2$? Explain.

TURN OVER

In this figure, AB is a diameter of circle O and chords AC and BD intersect at E so that $\angle CEB = 45°$.

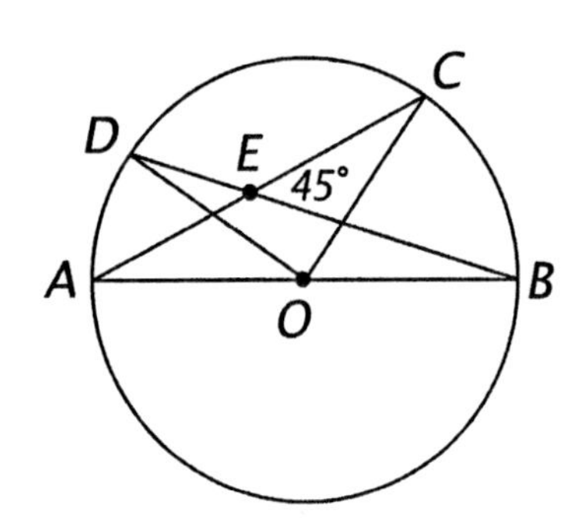

From this information, what can you conclude about

16. $m\overset{\frown}{AD} + m\overset{\frown}{CB}$?
17. $m\overset{\frown}{DC}$?
18. $\angle DOC$?
19. DO and OC?

In the figure below, D, E, and F are the midpoints of the sides of $\triangle ABC$ and X, Y, and Z are three of its centers.

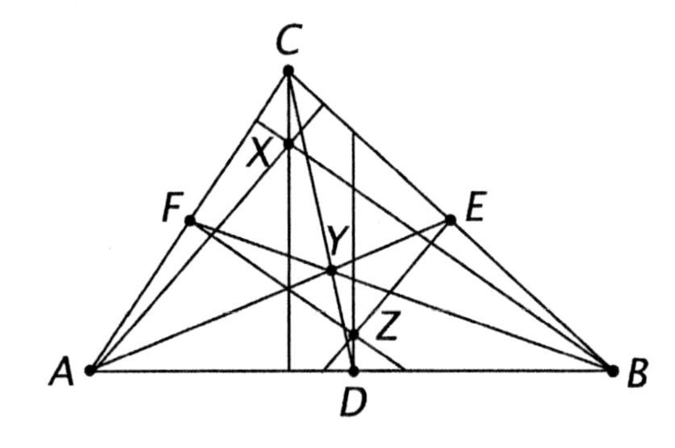

In which point are

20. the perpendicular bisectors of the sides concurrent?
21. the medians concurrent?

What center of $\triangle ABC$ is

22. point X?
23. point Z?

This figure shows the projection of a "hypercube" on a grid.

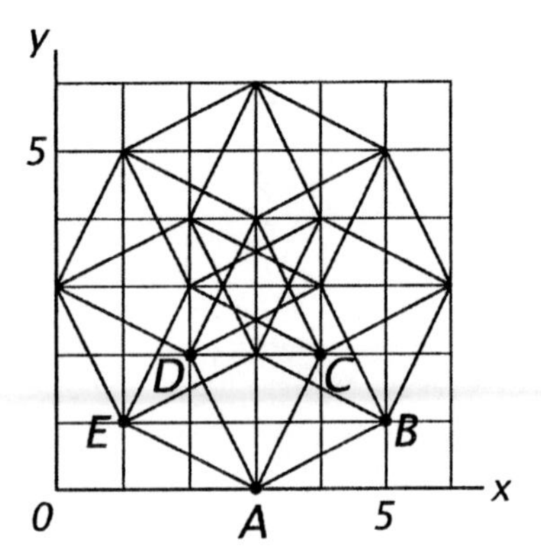

24. How long are AB, AC, AD, and AE?
25. Find the slopes of AB, AC, AD, and AE.
26. What do the slopes of AB and AD prove about them? Explain.
27. What can you conclude about the segments in the figure whose slopes are equal?

This figure appeared in a problem on an SAT exam.

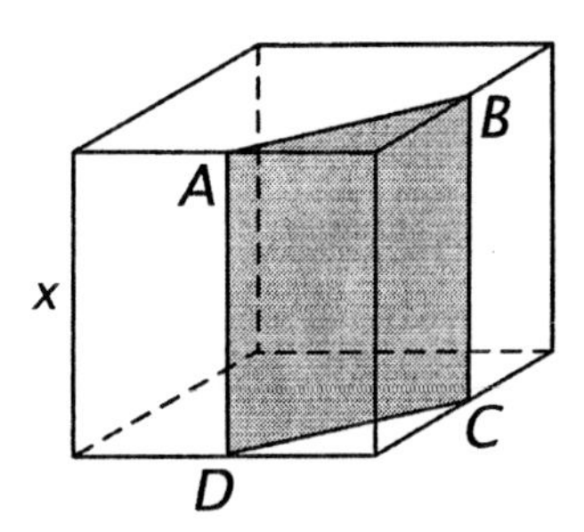

It shows a cube cut into two pieces by a slice that goes through the midpoints of four of its edges.

28. What kind of solids are the two pieces formed by the slice?

Given that the length of each edge of the cube is x, write an expression for

29. the volume of the smaller piece.
30. the area of the slice ABCD.

CONTINUED ON NEXT PAGE

This figure shows the path of a billiard ball bouncing around inside ΔABC.

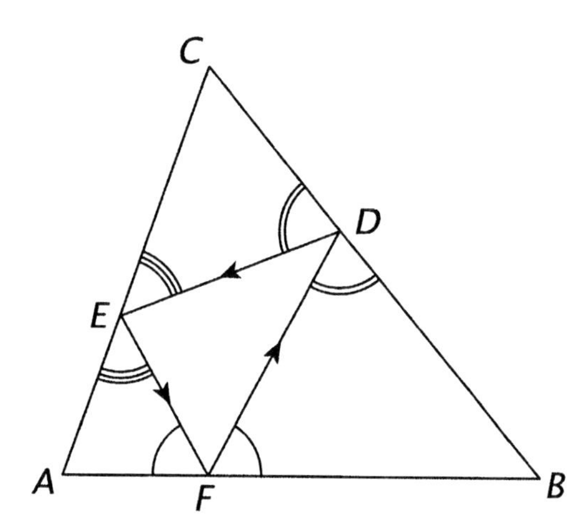

31. Given that the angles marked with identical arcs are equal and that $\angle A = 70°$ and $\angle DEF = 80°$, find the measures of the rest of the angles in the triangles and *write them on the figure on your answer sheet.*

32. Are any of the triangles in the figure similar? Explain.

To be able to drink through a straw, not only is the weight of the liquid important but also its attraction to the surface of the straw.

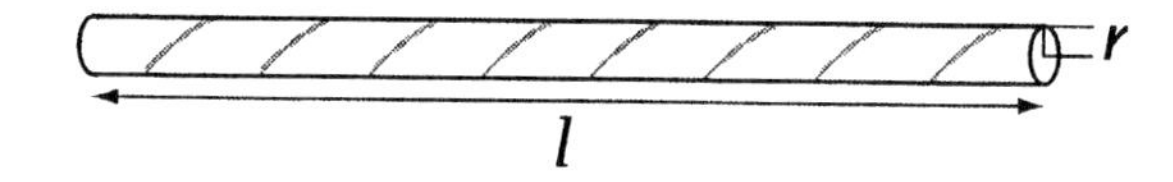

The weight of liquid in a straw depends on the volume of the straw.

33. Write an expression for the volume of a straw in terms of its length, l, and its radius, r.

The attraction of the liquid to the surface of the straw depends on the area of this surface.

34. Write an expression for the area of this surface in terms of its length, l, and its radius, r.

35. Write an expression for the ratio of this area to the volume of the straw and simplify it.

The bigger this ratio, the harder it is to drink through the straw.

36. Does the ratio depend on the length of the straw?

37. According to your expression for the ratio, what kind of straw would be hard to drink through?

This spiral-like curve consists of six 90° arcs of circles having radii of 1, 2, 3, 4, 5, and 6 units and centered on the corners of square ABCD.

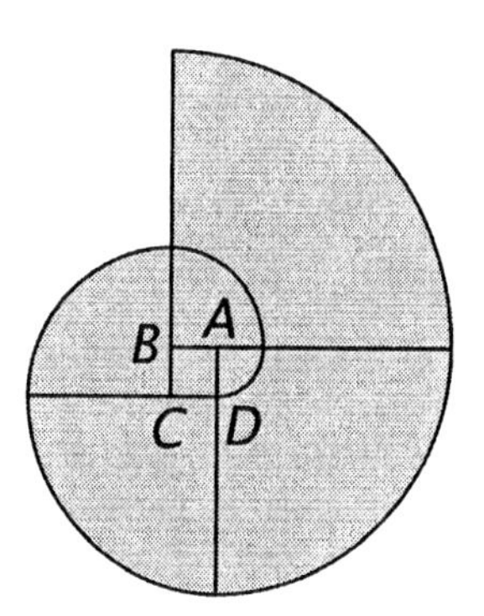

38. Find the exact length of the curve in terms of π.

39. Find its length to the nearest unit.

40. Find the exact area of the shaded region in terms of π.

GEOMETRY: *Answer Sheet for* Final Exam–Part 2

Name______________________________

1. ______ 4. ______ 7. ______

2. ______ 5. ______ 8. ______

3. ______ 6. ______

9. ______________________________

____

10. ______________________________

____

11. ______________________________

____

12. ______________________________

____

13. ______________________________

14. ______________________________

15. ______________________________

____

16. ______________________________

17. ______________________________

18. ______________________________

19. ______________________________

20. ______________________________

21. ______________________________

22. ______________________________

23. ______________________________

24. ______________________________

25. ______________________________

26. ______________________________

____

____

27. ______________________________

28. ______________________________

29. ______________________________

30. ______________________________

31.

C D E A F B

32. ______________________________

____

____

33. ______________________________

34. ______________________________

35. ______________________________

TURN OVER

36. ______________________________

37. ______________________________

____

38.

____

39. ______________________________

40.

____

GEOMETRY: *Answers to* Test on Chapter 1

1. Euclid.
2. A point.
3. A line extends without end in both directions. A line segment is part of a line bounded by two endpoints.
4. A rectangle is bounded by four line segments. A plane has no boundaries.
5. A polygon that has four sides.
6. A pentagon has five sides, so the pentathlon has five events.
7. An octopus has eight arms and an octagon has eight sides.
8. Triangles.
9. Three.
10. Coplanar.
11.

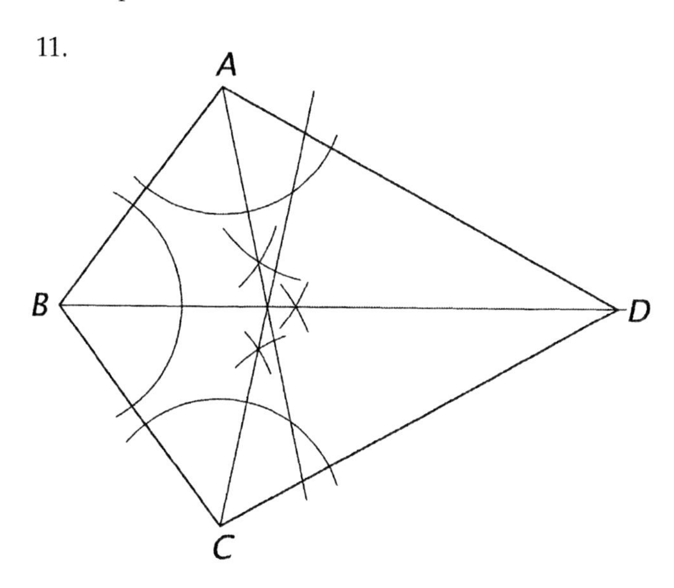

12. They seem to be concurrent.
13. It also appears to bisect $\angle D$.
14. The perimeter of room A is $2(15 \text{ ft}) + 2(18 \text{ ft}) = 66$ ft.
The perimeter of room B is $2(10 \text{ ft}) + 2(24 \text{ ft}) = 68$ ft.
Room B has the greater perimeter.
15. The area of room A is $(15 \text{ ft})(18 \text{ ft}) = 270$ sq ft.
The area of room B is $(10 \text{ ft})(24 \text{ ft}) = 240$ sq ft.
Room A has the greater area.
16. 7.3 cm.
17. 7.8 cm.
18. 62°.
19. 109°.
20. FB appears to be longer than FD, but it is shorter.
21.

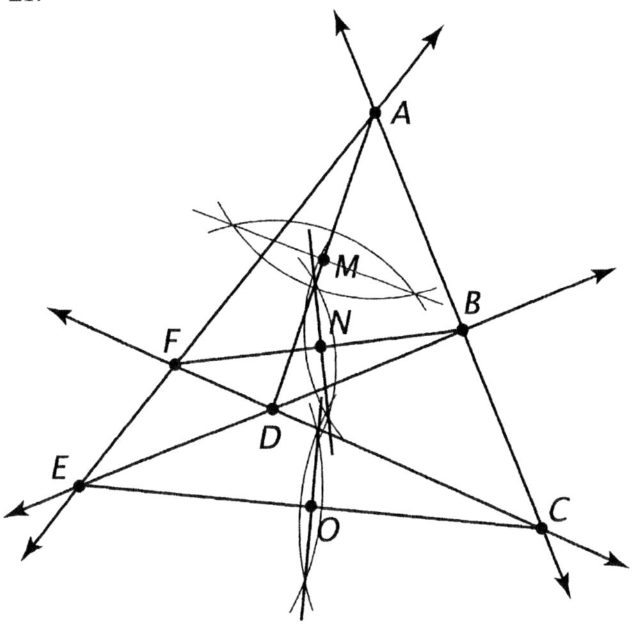

22. They seem to be collinear.
23. According to the formula used by the Egyptians for the area of a quadrilateral, the area of both quadrilaterals is $\frac{1}{4}(4 + 4)(5 + 5) = \frac{1}{4}(8)(10) =$ 20 square units.
This is correct for rectangle ABCD, whose area is 20 square units. If quadrilateral ABEF is cut by a vertical line through point F, the two pieces can be rearranged to form a square with sides of 4 units, so its area is actually 16 square units.

Extra Credit.

1. *(Student answer.)* Two of the cards are clearly the 2 and 3 of diamonds. The card in back could be the ace of diamonds.
2. Someone might think that the card in back is the ace of diamonds whereas it is actually the ace of hearts.

GEOMETRY: *Answers to* Test on Chapter 2

1. a) Associative property of multiplication.
   b) Commutative property of addition.

2. a) $(13 - 5)^2 = 8^2 = 64$.
   b) $13^2 - 5^2 = 169 - 25 = 144$.

3. a) $4x^2$.
   b) $6x - x - y + 7y = 5x + 6y$.

4. a) False.
   b) True.
   c) False.
   d) False.
   e) True.

5. a) An Euler diagram.
   b) If we don't have it, you don't want it.
   c) You don't want it if we don't have it.
   d) Statement 2.

6. a) If you watch Sesame Street, you love Eskimo pies.
   b) No.
   c) Yes.

7. a) If and only if.
   b) If you are an egotist, you are always me-deep in conversation
      If you are always me-deep in conversation, you are an egotist.
   c) One statement is the converse of the other.

8. a) Definition. A statement that gives the meaning of a word.
   b) Postulate. A statement assumed true without proof.
   c) Theorem. A statement proved by reasoning deductively.
   d) Euclid constructed the Geometry starting from definitions, postulates and axioms with which he demonstrated theorems.

9. a) True.
   b) True.
   c) True.
   d) False.
   e) True.

10. a) If the sky becomes dark, the crickets think that it is night.
    b) If the crickets start chirping, the temperature is estimated by counting cricket chirps.
    c) Direct.

11. a) The utensils are not stainless steel.
    b) The utensils are stainless steel.
    c) They will not rust.
    d) They are rusting.

12. a) $16^2 = 256$.
    b) $34^2 = 1{,}156$.
    c) $1{,}156 - 256 = 900$.
    d) $\sqrt{900} = 30$.

13. a) $2x = 80$, so $x = 40$.
    b) $\angle ABC = 180° - 80° - 30° = 70°$.
    c) $2y = 70$, so $y = 35$.
    d) $z = 180 - 40 - 35 = 105$.

14. a) $c = \pi d$, $36 = \pi d$, $d = \frac{36}{\pi} \approx 11.5$. 11.5 inches.
    b) 5(3 feet) = 15 feet (or 180 inches).

GEOMETRY: *Answers to* Test on Chapter 3

1. a) $9x - 72$.

   b) $16x + 4x^2$.

2. a) $3x + 20 = 8$,
   $3x = -12$,
   $x = -4$.

   b) $7(x - 4) = 5x + 2$,
   $7x - 28 = 5x + 2$,
   $2x = 30$,
   $x = 15$.

3. a) True.

   b) True.

   c) False.

   d) False.

   e) True.

4. a) Two angles are complementary iff they add to give 90°.

   b) Acute.

5. a) Obtuse.

   b) Acute.

   c) Right.

   d) ∠ADE and ∠EDC (or ∠DAE and ∠EAB).

   e) ∠EFC and ∠EFB.

6. a) 10.

   b) BF + FI = BI.

   c) ∠AHB = ∠BHC.

   d) GE = EC.

   e) ∠AEB = ∠IEH.

7. a) ∠ABD = 19.5°.

   b) ∠ABE = 160.5°.

   c) ∠ABD would get smaller.

   d) ∠ABE would get larger.

8. a) The <u>points</u> on a line can be numbered so that positive number <u>differences</u> measure <u>distances</u>.

   b) C is the midpoint of BD because it divides it into two equal segments: BC = CD = 3.

   c) Yes. C-D-F because $5 < 8 < 12$.

   d) DF.

   e) BD or CE.

   f) BE.

9. a) ∠HOM = 114 – 62 = 52°.

   b) 102. (62 + 40 = 102.)

   c) ∠TOM = 114 – 102 = 12°.

   d) ∠TOC = 24°.

   e) 126. (114 + 12 = 126.)

10. ∠AOB, ∠BOC, and ∠AOC are right angles because perpendicular lines form right angles. They are equal because all right angles are equal.

11. Because AB ⊥ BC and CD ⊥ BC, it follows that ∠ABC and ∠BCD are right angles, so ∠1 and ∠2 are complementary, as are ∠3 and ∠4. Because ∠2 = n°, ∠1 = $(90 - n)$°; because ∠3 = n°, ∠4 = $(90 - n)$°. In ΔBEC, ∠2 + ∠3 + ∠BEC = 180° because the sum of the angles of a triangle is 180°. So ∠BEC = 180° – ∠2 – ∠3 = 180° – n° – n° = $(180 - 2n)$° by subtraction and substitution. ∠1 + ∠4 = $(90 - n)$° + $(90 - n)$° = $(180 - 2n)$° by substitution, so ∠BEC = ∠1 + ∠4.

GEOMETRY: *Answers to* Test on Chapter 4

1. a) $2x^2 + 4x - 3$.
   b) $8x - 5y$.
   c) $5x^2 + 18x - 8$.
   d) $36x^2 + 12xy + y^2$.

2. a) AG = GD.
   b) If two angles of a triangle are equal, the sides opposite them are equal.

3. a) ∠B was copied as ∠D, BA and BC were copied on the sides of ∠D, EF was drawn.
   b) Yes. ΔEDF ≅ ΔABC by SAS.

4. a) SSS.
   b) Corresponding parts of congruent triangles are equal.
   c) A triangle is equilateral if all of its sides are equal.
   d) An equilateral triangle is equiangular.
   e) A triangle is isosceles if it has at least two equal sides.
   f) If two sides of a triangle are equal, the angles opposite them are equal.

5. a) ∠S = 45°. (Since EV ⊥ VS, ∠V is a right angle and so ∠V = 90°; ∠S = 180° – 90° – 45° = 45°.)
   b) ΔVES is a right triangle and it is also isosceles.
   c) $ES^2 = EV^2 + VS^2$.
   d) $93^2 = x^2 + x^2$; $2x^2 = 8{,}649$; $x^2 = 4{,}324.5$; $x \approx 66$. VS is approximately 66 million miles.

6. a) ΔAPC and ΔBPC are isosceles.
   b) ΔAPB ≅ ΔACB.
   c) Corresponding parts of congruent triangles are equal.
   d) SAS. (PB = BC, ∠PBA = ∠CBA, and BD = BD.)
   e) Corresponding parts of congruent triangles are equal.
   f) If the angles in a linear pair are equal, then their sides are perpendicular.

7. a)

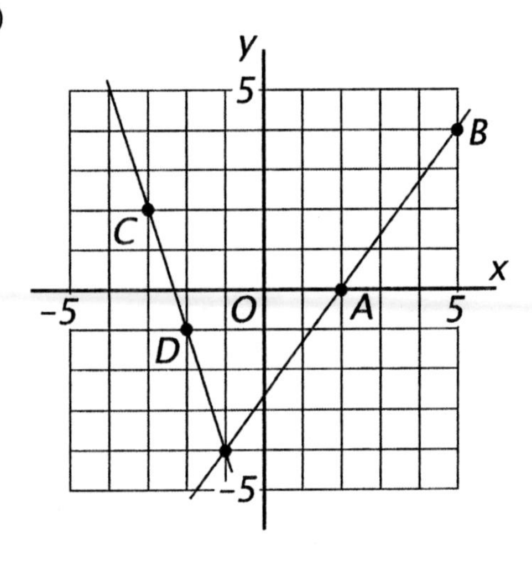

   b) $d = \sqrt{(x_2 - x_1)^2 + (y_2 - y_1)^2}$.
   c) $AB = \sqrt{(5-2)^2 + (4-0)^2} = \sqrt{3^2 + 4^2} = \sqrt{9 + 16} = \sqrt{25} = 5$.
   d) $CD = \sqrt{(-2 - -3)^2 + (-1-2)^2} = \sqrt{1^2 + (-3)^2} = \sqrt{1 + 9} = \sqrt{10}$.
   e) (–1, –4).

8.

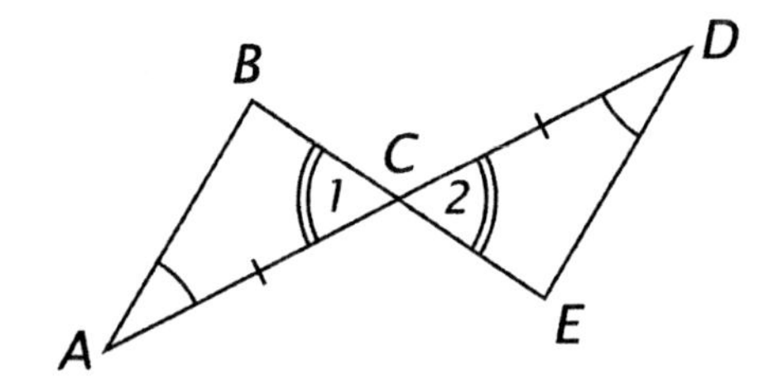

Given: C is the midpoint of AD; ∠1 and ∠2 are vertical angles; ∠A = ∠D.
Prove: AB = DE.

Proof.
1) C is the midpoint of AD. (Given.)
2) AC = CD. (The midpoint of a line segment divides it into two equal segments.)
3) ∠1 and ∠2 are vertical angles. (Given.)
4) ∠1 = ∠2. (Vertical angles are equal.)
5) ∠A = ∠D. (Given.)
6) ΔABC ≅ ΔDEC. (ASA.)
7) AB = DE. (Corresponding parts of congruent triangles are equal.)

GEOMETRY: *Answers to* Test on Chapter 5

1. a) $x(x + 8)$.

   b) $(x + 3)(x + 5)$.

   c) $(x + 3y)(x - 3y)$.

   d) $(5x - 4)(x + 2)$.

2. a) $<$.

   b) $<$.

   c) $>$.

3. a) $\angle CBD = 180° - 117° = 63°$.

   b) BC.

   c) CD.

4. a) True.

   b) False.

   c) False.

   d) True.

   e) True.

5. a) If a line bisects an angle, it divides it into two equal angles.

   b) An exterior angle of a triangle is greater than either remote interior angle.

   c) Substitution.

   d) If two angles of a triangle are unequal, the sides opposite them are unequal in the same order.

6. Yes. Since AB > CD, AB + BC > BC + CD (addition property of inequality). Since AC = AB + BC and BD = BC + CD (betweenness of points theorem), AC > BD by substitution.

7. Making a list of sets of three integers whose sum is 12 gives:

1, 1, 10	2, 2, 8	3, 3, 6	4, 4, 4
1, 2, 9	2, 3, 7	3, 4, 5	
1, 3, 8	2, 4, 6		
1, 4, 7	2, 5, 5		
1, 5, 6			

   Comparing the numbers in these sets with the fact that the sum of any two sides of a triangle is greater than the third side (the Triangle Inequality Theorem) reveals that the only other such triangles have sides 2, 5, 5 and 3, 4, 5.

GEOMETRY: *Answers to* Test on Chapter 6

1. a) $\frac{3}{x^2}$.

   b) $\frac{5(x-1)}{10(x-1)} = \frac{1}{2}$.

   c) $\frac{-1}{x-4}$.

   d) $\frac{2y}{xy}$ and $\frac{8x}{xy}$.

2. a) True.

   b) True.

   c) False.

   d) False.

   e) True.

3. In a plane, two points each equidistant from the endpoints of a line segment determine the perpendicular bisector of the line segment.

4. a) 110°.

   b) 70°.

   c) 55°.

5. a)

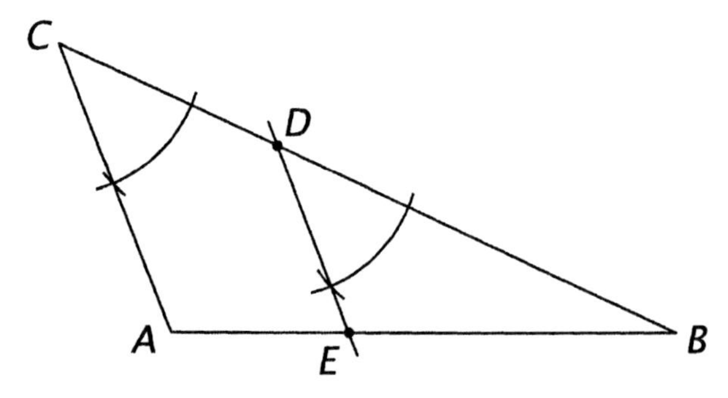

   b) ∠A and ∠AED are supplementary.

   c) Parallel lines form supplementary interior angles on the same side of a transversal.

6. a) ∠ABD = ∠CDF.

   b) Parallel lines form equal corresponding angles.

   c) ∠FAB and ∠B are complementary.

   d) The acute angles of a right triangle are complementary.

   e) EA || FB.

   f) Equal alternate interior angles mean that lines are parallel.

7. a) Vertical angles are equal.

   b) If two angles of one triangle are equal to two angles of another triangle, the third angles are equal.

   c) ASA (or AAS).

   d) Corresponding parts of congruent triangles are equal.

   e) If two sides of a triangle are equal, the angles opposite them are equal.

8. Since AD ⊥ BC, ∠ADB and ∠ADC are right angles, so ΔADB and ΔADC are right triangles. The acute angles of a right triangle are complementary, so ∠1 and ∠B are complementary and ∠2 and ∠C are complementary. From this we know that ∠1 + ∠B = 90° = ∠2 + ∠C, so ∠1 – ∠2 = ∠C – ∠B by subtraction.

GEOMETRY: *Answers to* Test on Chapter 7

1. a) $\frac{5x}{10} + \frac{2x}{10} = \frac{7x}{10}$.

   b) $\frac{2(x+6)}{4 \cdot 3} = \frac{x+6}{6}$.

   c) $\frac{1}{7} \cdot \frac{x}{7} = \frac{x}{49}$.

   d) $\frac{3(x+y)}{(x+y)(x-y)} - \frac{1}{x-y} = \frac{3}{x-y} - \frac{1}{x-y} = \frac{2}{x-y}$.

2. A *rhombus* is a quadrilateral all of whose sides are equal.

3. a) False.

   b) True.

   c) False.

   d) True.

   e) False.

4. a) A quadrilateral is a parallelogram if its opposite sides are equal.

   b) The opposite sides of a parallelogram are parallel.

5. a) $\angle P = (180 - x - y)°$.

   b) $\angle A = (180 - 2x)°$.

   c) $\angle B = (180 - 2y)°$.

   d) $\angle A + \angle B = (360 - 2x - 2y)°$.

   e) $\angle P = \frac{1}{2}(\angle A + \angle B)$.

6. a) Parallelograms.

   b) Trapezoids.

   c) By the Midsegment Theorem. GH is a midsegment of ΔADC and EF is a midsegment of ΔABC. A midsegment of a triangle is parallel to the third side.

   d) From the fact that E, F, G, and H are the midpoints of the sides of square ABCD, we can conclude that DG $\parallel$ EB and DG = EB. It follows that DGBE is a parallelogram (a quadrilateral is a parallelogram if two opposite sides are both parallel and equal), so DE $\parallel$ GB (the opposite sides of a parallelogram are parallel).

GEOMETRY: *Answers to* Test on Chapter 8

1. a) $\sqrt{99} = \sqrt{9 \cdot 11} = \sqrt{9}\sqrt{11} = 3\sqrt{11}$.

   b) $\sqrt{72} + \sqrt{2} = \sqrt{36 \cdot 2} + \sqrt{2} = \sqrt{36}\sqrt{2} + \sqrt{2} = 6\sqrt{2} + \sqrt{2} = 7\sqrt{2}$.

   c) $(8\sqrt{3})^2 = 8^2(\sqrt{3})^2 = 64 \cdot 3 = 192$.

   d) $\sqrt{17^2 - 15^2} = \sqrt{289 - 225} = \sqrt{64} = 8$.

2. a) True.

   b) True.

   c) False.

   d) False.

   e) False.

3.

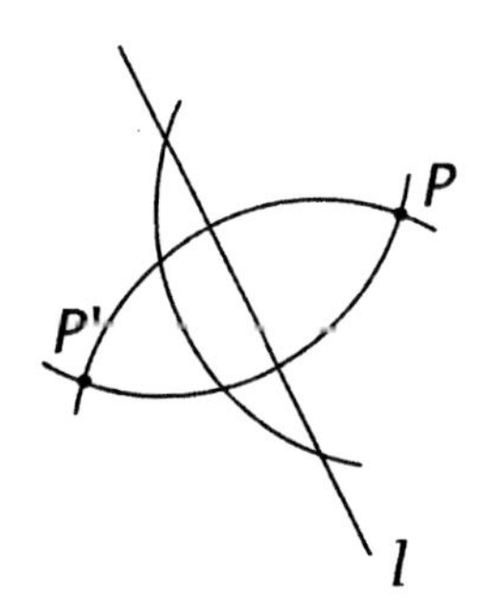

4. a) Glide reflection.

   b) Translation.

   c) Rotation.

   d) Yes.

   e) If two figures are congruent, there is an isometry such that one figure is the image of the other.

5. a)

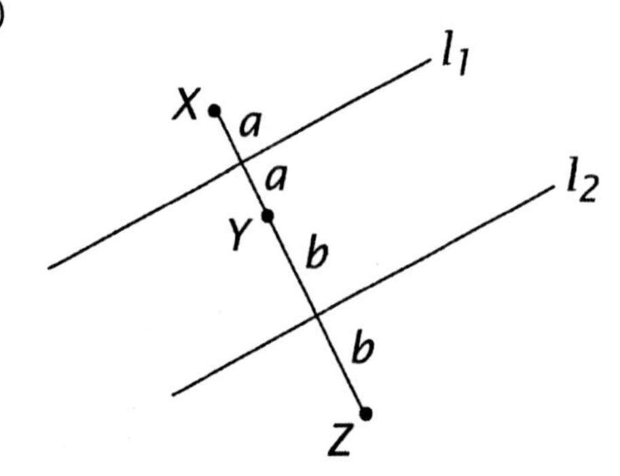

   b) $a + b$.

   c) $2a + 2b$.

   d) A translation.

   e) It is twice the distance between l_1 and l_2.

6. a) Rotation symmetry.

   b) 5.

   c) 72°. ($\frac{360°}{5}$.)

   d) Yes. $3 \times 72° = 216°$.

   e) No. If it were rotated 180°, it would not coincide with itself.

7. a) Yes. The board has two lines of symmetry which contain its diagonals.

   b) Yes. It has rotation symmetry because it can be rotated 180° to coincide with itself.

8. a)

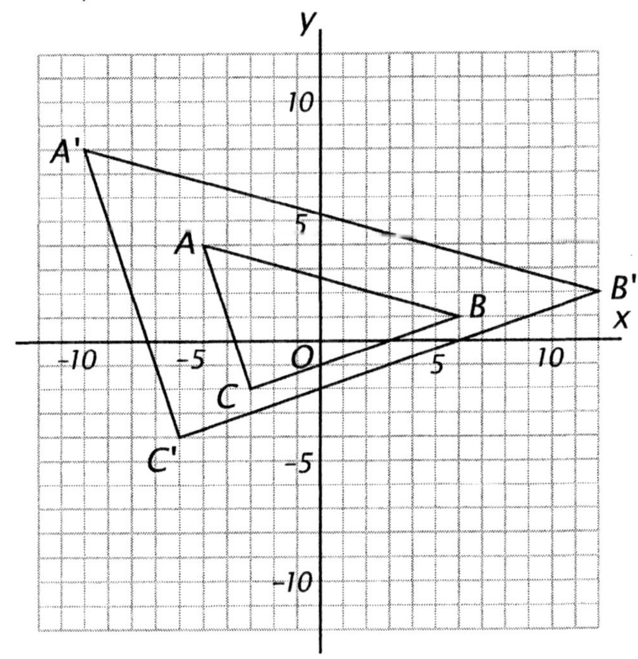

   b) They are twice as large.

   c) A dilation.

   d) No. Distance is not preserved.

GEOMETRY: *Answers to* Test on Chapter 9

1. a) False.

   b) True.

   c) True.

   d) True.

   e) False.

2. a) $x = \sqrt{256} = 16$.

   b) $y = \sqrt{324} = 18$.

   c) $Z = 24^2 = 576$.

   d) No. The square of the longest side is not equal to the sum of the squares of the other two sides.

3. a) αΔADG + αDEFG + αΔEBF + αΔGFC = 20 + 48 + 20 + 15 = 103.

   b) $\frac{1}{2}16 \cdot 13 = 8 \cdot 13 = 104$.

   c) According to the Area Postulate, the area of a polygonal region is equal to the sum of the areas of its nonoverlapping parts, but $104 \neq 103$.

   d) *(Extra credit if answered correctly)* Although ABC looks like a triangle, it is actually a pentagon.

4. a) 660 meters.

   b) 22,200 square meters.

5. a) One method: Divide the figure into four trapezoids and a square. Its area is $4(\frac{1}{2}2(1 + 3)) + 1 = 16 + 1 = 17$ square units.

   b) $25 - 17 = 8$; $\frac{8}{4} = 2$ square units.

6. a) The midpoint of a line segment divides it into two equal segments.

   b) Vertical angles are equal.

   c) All right angles are equal.

   d) AAS.

   e) Congruent triangles have equal areas.

   f) αEFGH = $\frac{1}{5}$αABCD.

7. One answer: The area of the figure is $12^2 = 144$. Since the areas of the two smaller squares are 36 and 25, the sum of the areas of A and B is $144 - 36 - 25 = 83$. The region labeled B is a right triangle with legs 12 and $12 - 5 = 7$, so its area is $\frac{1}{2}12 \cdot 7 = 42$. So the area of the region labeled A is $83 - 42 = 41$. The area of region B is slighter more than the area of region A.

GEOMETRY: *Answers to* Test on Chapter 10

1. a) $12 = 30x$, $x = \frac{12}{30} = \frac{2}{5}$ or 0.4.

   b) $3(x - 5) = x$, $3x - 15 = x$, $2x = 15$, $x = \frac{15}{2}$ or 7.5.

   c) $5x = 3(x + 6)$, $5x = 3x + 18$, $2x = 18$, $x = 9$.

   d) $2x + x = 48$, $3x = 48$, $x = 16$.

2. a) True.

   b) True.

   c) False.

   d) False.

   e) True.

3. a) A proportion.

   b) $h = \frac{3}{2}w$.

   c) $h = \frac{3}{2}60 = 90$. 90 feet high.

4. a) $\frac{x}{18} = \frac{7}{12}$, $12x = 126$, $x = 10.5$.

   b) $\frac{x}{9} = \frac{6}{10}$, $10x = 54$, $x = 5.4$.

   c) $\frac{x}{18} = \frac{2}{x}$, $x^2 = 36$, $x = 6$.

5. a) $\frac{3}{5}$ or 0.6.

   b) $(\frac{3}{5})^2 = \frac{9}{25}$ or 0.36.

   c) $\frac{45}{x} = \frac{3}{5}$, $x = 75$.

   d) $\frac{36}{x} = \frac{9}{25}$, $x = 100$.

6. a) If a line parallel to one side of a triangle intersects the other two sides in different points, it divides the sides in the same ratio.

   b) Substitution.

   c) $\frac{PS}{100} = \frac{100}{15}$, $PS = \frac{10{,}000}{15} \approx 667$ pu.

7. a) Corresponding angles of similar triangles are equal.

   b) Corresponding angles of similar triangles are equal.

   c) ΔCAD and ΔADB are isosceles.

8. By applying the Pythagorean Theorem to ΔADE, $AE = 75$. Since $EF \parallel AB$, $\angle CEF = \angle A$ and $\angle CFE = \angle B$, so all four triangles in the figure are similar by AA. Using the fact that corresponding sides of similar triangles are proportional gives us the lengths of the other line segments. They are marked in the figure below.

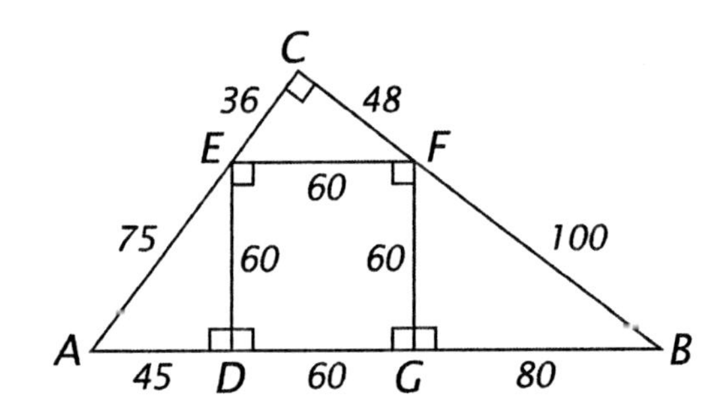

GEOMETRY: *Answers to* Test on Chapter 11

1. a) $x^2 - 4x - 12 = 0$, $(x - 6)(x + 2) = 0$, $x - 6 = 0$ or $x + 2 = 0$, $x = 6$ or $x = -2$. The solutions are 6 and –2.

   b) $b^2 = c^2 - a^2$, $b = \sqrt{c^2 - a^2}$.

   c) $\frac{9}{5}C = F - 32$, $F = \frac{9}{5}C + 32$.

2. a) $\frac{x}{3} = \frac{3}{5}$, $5x = 9$, $x = \frac{9}{5}$ or 1.8.

   b) $x = 12$, $y = 6\sqrt{3}$.

   c) $x = 8\sqrt{2}$.

3. a) $a = \sqrt{2}$; $b = \sqrt{3}$; $c = \sqrt{4} = 2$; $d = \sqrt{5}$; $e = \sqrt{6}$; $f = \sqrt{7}$; $g = \sqrt{8}$ (or $2\sqrt{2}$); $h = \sqrt{9} = 3$; $i = \sqrt{10}$; $j = \sqrt{11}$; $k = \sqrt{12}$ (or $2\sqrt{3}$); $l = \sqrt{13}$.

   b) $x = 45°$; $\tan y = \frac{2}{1} = 2$, $y \approx 63°$; $\tan z = \frac{3}{1} = 3$, $z \approx 72°$.

4. a) $AM = \frac{1}{2}(100)(12) = 600$ in.;

   $AC = 600 + \frac{1}{2}(0.5) = 600.25$ in.

   b) $600^2 + MC^2 = 600.25^2$, $MC^2 = 600.25^2 - 600^2$,

   $MC = \sqrt{600.25^2 - 600^2} \approx 17$ in.

5. a) $\angle POA = 30°$, $\angle APO = 60°$ and $\angle PAO = 90°$.

   b) $PA = \frac{1}{2}$, $AO = \frac{1}{2}\sqrt{3}$.

   c) $(-\frac{1}{2}\sqrt{3}, \frac{1}{2})$.

6. a) $m = -\frac{8}{10} = -0.8$ or $-\frac{4}{5}$.

   b) $y = -\frac{1}{3}x + 8$.

   c) $0 = -\frac{1}{3}x + 8$, $\frac{1}{3}x = 8$, $x = 24$; (24, 0).

7. a) $AM = 20$ cm.

   b) $\sin \angle APM = \frac{20}{70}$; $\angle APM \approx 16.6°$.

   c) $\angle APB = 2\angle APM \approx 33°$.

8. $\angle C = 180° - 85° - 65° = 30°$. By the Law of Sines, $\frac{\sin 30°}{400 \text{ m}} = \frac{\sin 65°}{AC}$, so $AC \sin 30° = (400 \text{ m})\sin 65°$ and $AC = \frac{\sin 65°}{\sin 30°}(400 \text{ m}) \approx 725$ m.

9. Each leg of a right triangle is the geometric mean between the hypotenuse and its projection on the hypotenuse, so $\frac{x}{a} = \frac{a}{c}$ and $\frac{y}{b} = \frac{b}{c}$. Multiplication gives us the equations $xc = a^2$ and $yc = b^2$. Dividing, we get $\frac{xc}{yc} = \frac{a^2}{b^2}$, so $\frac{x}{y} = \frac{a^2}{b^2}$.

GEOMETRY: *Answers to* Test on Chapter 12

1. a)

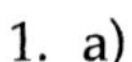

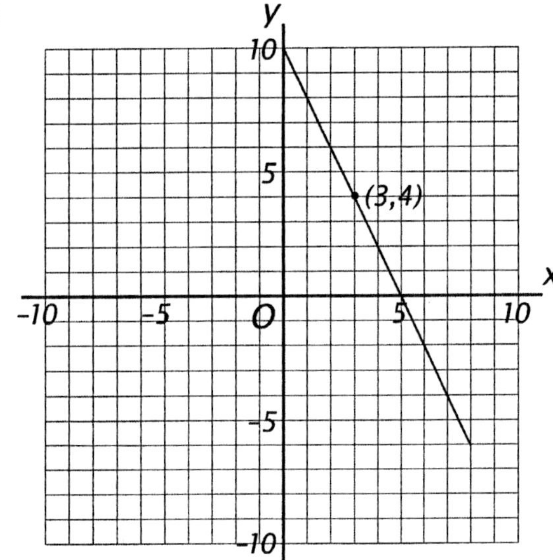

b) $y - 4 = -2(x - 3)$.

c) $y - 4 = -2x + 6$, $y = -2x + 10$.

d) 10.

2. a) False.

b) True.

c) False.

d) True.

e) True.

3. a) $12x = 18(20)$, $12x = 360$, $x = 30$.

b) $x = \frac{1}{2}(80° + 110°) = \frac{1}{2}(190°) = 95°$.

c) $x = \frac{1}{2}(95° - 35°) = \frac{1}{2}(60°) = 30°$.

4. a) (One solution)

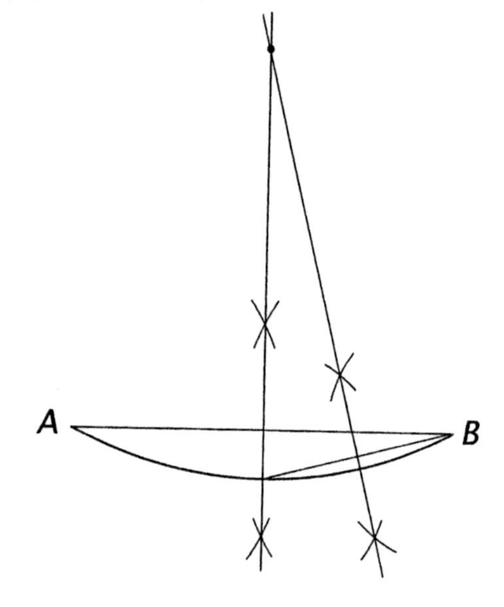

b) The perpendicular bisector of a chord of a circle contains the center of the circle.

5. a) $m\widehat{AB} = \frac{360°}{15} = 24°$.

b) $\angle C = \frac{1}{2}(24°) = 12°$.

c) $15 \times 12° = 180°$.

6. a) An angle inscribed in a semicircle is a right angle.

b) An angle inscribed in a quarter of a circle intercepts an arc that is three quarters of the circle, or 270°. An inscribed angle is equal in measure to half its intercepted arc, so an angle inscribed in a quarter of a circle has a measure of 135°.

7.

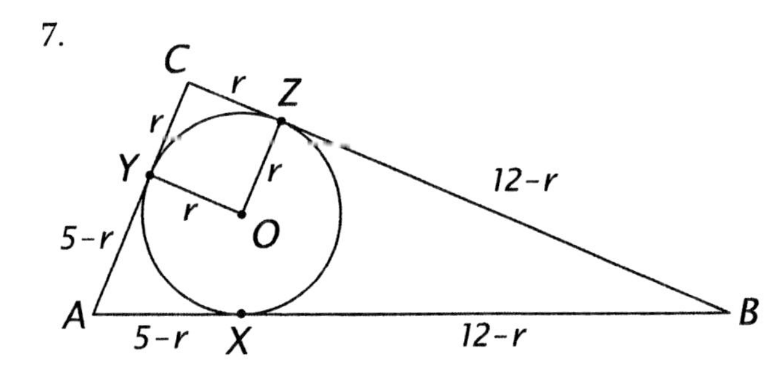

a) ΔABC is a right triangle because $5^2 + 12^2 = 13^2$.

b) OYCZ is a square.

c) $(5 - r) + (12 - r) = 13$, $17 - 2r = 13$, $-2r = -4$, $r = 2$.

8. a) If a line is tangent to a circle, it is perpendicular to the radius drawn to the point of contact.

b) Since ΔABC is a right triangle, $AB^2 + BC^2 = AC^2$, so $AB^2 + 2100^2 = 2115^2$, $AB^2 = 2115^2 - 2100^2 = 63{,}225$, $AB = \sqrt{63{,}225} \approx 251$. You could see approximately 250 miles to the horizon.

9. $\Delta AEB \sim \Delta CED$ by AA ($\angle A = \angle C$ and $\angle B = \angle D$ because inscribed angles that intercept the same arc are equal). So $\frac{AB}{CD} = \frac{AE}{CE}$ because corresponding sides of similar triangles are proportional.

GEOMETRY: *Answers to* Test on Chapter 13

1. a) Substituting $3y$ for x in the second equation gives $5(3y) - 6y = 36$, $15y - 6y = 36$, $9y = 36$, $y = 4$. Substituting this value for y in the first equation gives $x = 3(4) = 12$. The solution to the system is (12, 4).

   b) Multiplying the first equation by 5 gives $20x + 5y = 90$. Subtracting the second equation gives $13x = 39$, so $x = 3$. Substituting this value for x in the first equation gives $4(3) + y = 18$, $12 + y = 18$, $y = 6$. The solution to the system is (3, 6).

2. a) True.

   b) True.

   c) False.

   d) True.

   e) False.

3. a) Concurrent lines are lines that contain the same point.

   b) $\frac{6}{7} \cdot \frac{5}{3} \cdot \frac{x}{8} = 1$, $30x = 168$, $x = 5.6$.

4. a)

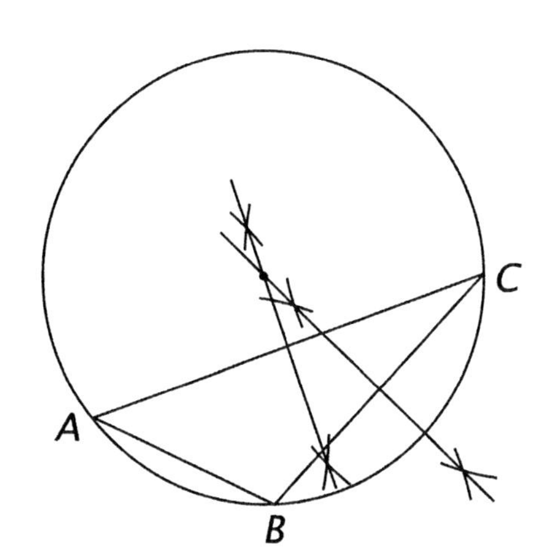

   b)

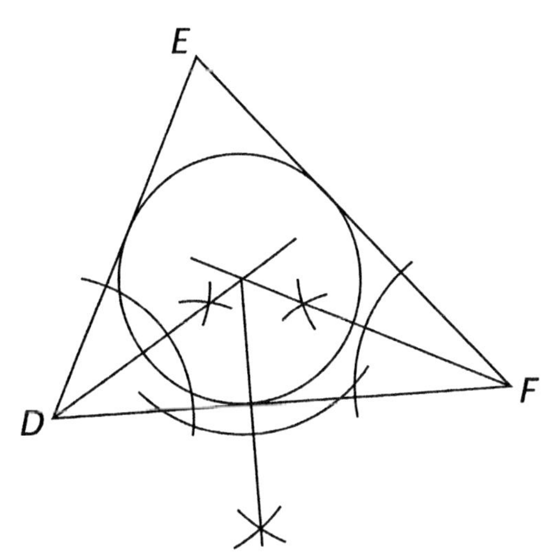

5. The geometric words are "altitudes" and "orthocenter" and his daughter's name is "Rachel."

6. a) It means that there is a circle that contains all of the vertices of the polygon.

   b) They are supplementary.

   c) The point in which its medians are concurrent.

   d) It is similar to it (or, it is a dilation of it).

7. a) Since ΔBCD is equilateral, each of its angles is 60° and intercepts an arc of 120°. So $\angle BPD = \frac{1}{2} m\widehat{BD} = \frac{1}{2} 120° = 60°$.

   b) $\angle CPD = \frac{1}{2} m\widehat{CD} = \frac{1}{2} 120° = 60°$.

   c) ∠BPA and ∠BPD are a linear pair, so they are supplementary; $\angle BPA = 180° - 60° = 120°$.

   d) ∠CPA and ∠CPD are a linear pair, so they are supplementary; $\angle CPA = 180° - 60° = 120°$.

   e) $\angle BPC = \angle BPD + \angle CPD = 60° + 60° = 120°$.

   f) The three lines from the Steiner Point to the vertices of the triangle form equal angles with each other.

8. a) The lines containing the altitudes of a triangle are concurrent.

   b) Its angle bisectors.

   c) $\angle 1 = \angle 2$. $\angle BEC = \angle BEA$ because $BE \perp AC$ and $\angle BED = \angle BEF$ because EB bisects ∠DEF, so $\angle 1 = \angle 2$ by subtraction.

   d) It would move along the sides of ΔDEF.

GEOMETRY: *Answers to* Test on Chapter 14

1. a) True.

   b) False.

   c) True.

   d) False.

   e) False.

2. a) $m\widehat{AB} = \frac{360°}{9} = 40°$.

   b) $m\widehat{BD} = 2(40°) = 80°$.

   c) $\angle BAD = \frac{1}{2} m\widehat{BD} = \frac{1}{2} 80° = 40°$.

   d) $\angle B = \frac{1}{2} m$ major $\widehat{AC} = \frac{1}{2} 7(40°) = 140°$.

   e) BC || AD because ∠BAD + ∠B = 180°. Supplementary interior angles on the same side of a transversal mean than lines are parallel.

   f) ABCD is an isosceles trapezoid. AB = CD because a regular polygon is equilateral and BC || AD.

   g) ADEF is a cyclic quadrilateral because its vertices lie on a circle. (Also, its opposite angles are supplementary.)

3. a) $\frac{1}{4}\pi x^2 = \frac{1}{2}(308)(204)$,
   $\pi x^2 = 2(308)(204) = 125{,}664$, $x^2 \approx 40{,}000$,
   $x \approx 200$.
   About 200 feet.

   b) $\frac{1}{4} 2\pi(200) = 100\pi \approx 314$.
   About 314 feet.

4. $16(\frac{1}{4}\pi 1^2 - \frac{1}{2}1^2) = 16(\frac{\pi}{4} - \frac{1}{2}) = 4\pi - 8$.

5. $\frac{\pi r^2}{(2r)^2} = \frac{\pi}{4} = \frac{11}{14}$, $\pi = \frac{44}{14} = \frac{22}{7}$ or $3\frac{1}{7}$.

6. a) Area $= Mr^2 = (8 \sin\frac{180}{8} \cos\frac{180}{8})(35)^2 \approx 3{,}465$.
   The area is approximately 3,465 square feet.

   b) Perimeter $= 2Nr = 2(8 \sin\frac{180}{8})(70) \approx 429$.
   The perimeter is approximately 429 feet.

   c) Area $= MR^2 - Mr^2 =$
   $(8 \sin\frac{180}{8} \cos\frac{180}{8})(70)^2 - 3{,}465 \approx$
   $13{,}859 - 3{,}465 \approx 10{,}394$.
   The area is approximately 10,400 square feet.

7. a) $\angle AOB = \frac{360°}{12} = 30°$.

   b) $\angle OAC = 60°$.

   c) $AC = \frac{1}{2}r$.

   d) $\alpha\Delta AOB = \frac{1}{2}(r)(\frac{1}{2}r) = \frac{1}{4}r^2$.

   e) $12(\frac{1}{4}r^2) = 3r^2$.

   f) The formula, $A = 3r^2$, is like the formula, $A = \pi r^2$. This is not surprising because a regular dodecagon looks a lot like its circumscribed circle.

GEOMETRY: *Answers to* Test on Chapter 15

1. a) False.
   b) True.
   c) True.
   d) True.
   e) False.

2. a) (400 m)(1,200 m) = 480,000 m^2.
   b) (400 m)(1,200 m)(200 m) = 96,000,000 m^3.
   c) 2(480,000 + 80,000 + 240,000) = 1,600,000 m^2.

3. a) $\pi r^2 h$.
   b) Vase A because of Cavalieri's Principle. Corresponding cross sections of vases A and B appear to have equal areas, suggesting that the vases have equal volumes.

4. a) $AB^2 = 9^2 + 3^2 = 81 + 9 = 90$; $AB = \sqrt{90}$ or $3\sqrt{10}$ cm.
   b) $\tan \angle BAO = \frac{3}{9}$, $\angle BAO \approx 18°$.
   c) $V = \frac{1}{3}\pi(3)^2(9) = 27\pi \approx 85$ cm^3.

5. a) $4\pi r^2 = 2{,}463$, $r^2 \approx 196$, $r \approx 14$. Its radius was about 14 inches.
   b) $V = \frac{4}{3}\pi(14)^3 \approx 11{,}494$. Its volume was about 11,494 cubic inches.

6. a) b^2 represents the area of the base.
   b) a^2 represents the area of the top face.
   c) $\frac{1}{3}b^2h$.
   d) A square pyramid.
   e) $\frac{1}{3}(a^2 + a^2 + a^2)h = \frac{1}{3}(3a^2)h = a^2h$.
   f) A square prism.

7. a) $\frac{40}{16} = 2.5$.
   b) $2.5^2 = 6.25$.
   c) $2.5^3 = 15.625$.
   d) (6.25)(60 ft^2) = 375 ft^2.
   e) (15.625)(400 lb) = 6,250 lb.
   f) 2.5(10 ft) = 25 ft.

GEOMETRY: *Answers to* Test on Chapter 16

1. a) True.
   b) True.
   c) True.
   d) True.
   e) False.

2. a) They are complementary.
   b) No. In right ΔAFG, for example, the acute angles are each 60° and their sum is 120°.
   c) That the quadrilateral is a parallelogram.
   d) No. There are no parallel lines in sphere geometry.

3. a) $\angle 2 + \angle B + \angle C < 180°$.
   b) $\angle 1 + \angle 2 = 180°$.
   c) $\angle 2 + \angle B + \angle C < \angle 1 + \angle 2$ (substitution), so $\angle B + \angle C < \angle 1$ (subtraction).
   d) In Lobachevskian geometry, an exterior angle of a triangle is greater than the sum of the remote interior angles.

4. a) A Saccheri quadrilateral.
   b) The line segment connecting the midpoints of the base and summit of a Saccheri quadrilateral is perpendicular to both of them.
   c) Birectangular quadrilaterals.
   d) $\angle F$ and $\angle D$ are acute. In Lobachevskian geometry, the summit angles of a Saccheri quadrilateral are acute.
   e) The net is less than 36 feet long. If the summit angles of a birectangular quadrilateral are unequal, the legs opposite them are unequal in the same order.

5. a) Ruler Postulate.
   b) A birectangular quadrilateral whose legs are equal is a Saccheri quadrilateral.
   c) The summit angles of a Saccheri quadrilateral are equal.
   d) An exterior angle of a triangle is greater than either remote interior angle.
   e) Substitution.
   f) Betweenness of Rays Theorem.
   g) "Whole Greater than Part" Theorem.
   h) Transitive.

6. a) The perimeter of ΔABC is half the perimeter of ΔDEF. In Euclidean geometry, a midsegment of a triangle is half as long as the third side.
   b) The perimeter of ΔABC is less than half the perimeter of ΔDEF.
   c) The sides of ΔABC are parallel to the sides of ΔDEF.
   d) No. There are no parallel lines in Riemannian geometry.
   e) No. In Lobachevskian geometry, if two triangles are similar, they must also be congruent. ΔABC and ΔDEF are not congruent.

GEOMETRY: *Answer Sheet for* Midyear Exam–Part 1

Name________________________________

1. $3x$
2. $4x^2 - x^3$
3. -11
4. $8x + 2y$
5. $(5x - 1)(x + 3)$
6. $\frac{1}{x+3}$
7. $\frac{x+4}{8}$
8. $6\sqrt{2}$
9. *The circumference of a circle is π times its diameter.*
10. *The area of a square is the square of its side.*
11. *If an animal is a porcupine, then it has long quills.*
12. *No.*
13. *It is the converse.*
14. *d*
15. *b*
16. *c*
17. *c*
18. *a*
19. *c*
20. *a*
21. *c*
22. *d*
23. *d*
24. *acute*
25. *True*
26. *True*
27. *definition*
28. *differences*
29. *perpendicular*
30. *True*
31. *y-axis*
32. *five*
33. *True*
34. *equal*
35. $AB + BC = AC$
36. *True*
37. *isosceles*
38. *may*
39. *complement*
40. *True*
41. $c > a$ and $c > b$
42. *True*
43. *four*

44/45.

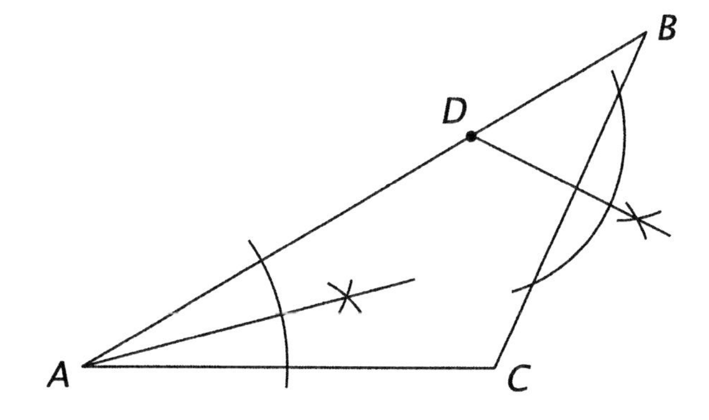

GEOMETRY: *Answer Sheet for* Midyear Exam–Part 2

Name______________________________

1. *The side opposite the right angle in a right triangle.*
2. *The "if-clause" of a conditional statement.*
3. *Line symmetry.*
4. *(One answer.) Fold the figure to see if the two halves coincide.*
5. *68°*
6. *34°*
7. *121*
8. *56°*
9. *31*
10. *<*
11. *=*
12. *=*
13. *<*
14. *>*
15. *ΔABC is isosceles.*
16. *A midsegment of a triangle is parallel to the third side.*
17. *Parallel lines form equal corresponding angles.*
18. *ABED is an isosceles trapezoid.*
19. *The base angles of an isosceles trapezoid are equal.*
20. *ADCE is a parallelogram (rectangle).*
21. *ΔBCE is an isosceles right triangle.*
22. *The opposite sides of a parallelogram are equal.*
23. *40 ft.*
24. *Yes. Corresponding sides of congruent polygons are equal.*
25. *Yes. For example, FE || HI because equal alternate interior angles mean that lines are parallel.*
26. *The sum of the angles of a quadrilateral is 360°.*
27. *Yes. AM = PM and BM = PM (corresponding parts of congruent triangles are equal), so AM = BM.*
28. *Yes. $\angle MPC = \angle A = 90°$ and $\angle MPD = \angle B = 90°$, so $\angle MPC + \angle MPD = 180°$, so $\angle CPD$ is a straight angle.*
29. *Yes. $\angle AMC = \angle PMC$ and $\angle BMD = \angle PMD$, so $\angle CMD = \frac{1}{2}\angle AMB = 90°$.*
30.

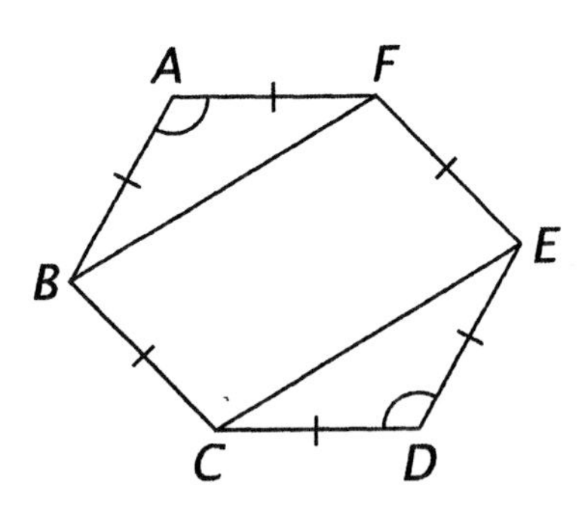

Given: AB = BC = CD = DE = EF = FA and $\angle A = \angle D$.
Prove: BC || FE.

1. *AB = BC = CD = DE = EF = FA and $\angle A = \angle D$ (Given.)*
2. *$\Delta ABF \cong \Delta DEC$. (SAS.)*
3. *BF = EC. (Corresponding parts of congruent triangles are equal.)*
4. *BCEF is a parallelogram. (A quadrilateral is a parallelogram if its opposite sides are equal.)*
5. *BC || FE. (The opposite sides of a parallelogram are parallel.)*

GEOMETRY: *Answer Sheet for* Final Exam–Part 1

Name________________________________

1. *True*
2. *volumes*
3. *tangent*
4. *areas*
5. *apothem*
6. *sphere*
7. *True*
8. *True*
9. *True*
10. *rectangle*
11. *diameter*
12. *shorter*
13. *altitudes*
14. *concurrent*
15. *hypotenuse*
16. *surface area*
17. *The base angles of an isosceles trapezoid are equal.*
18. *AAS*
19. *18*
20. *24*
21. *768*
22. *To be equal to the height of the post.*
23. *Isosceles right triangles.*
24. *That they are parallel.*
25. *They are similar.*
26. *By the length of the cliff's shadow.*
27. *The Intersecting Chords Theorem.*
28. *That rectangles having the segments of the chords as their sides have equal areas.*
29. *Squares.*
30. $\tan 23^\circ = \frac{AC}{7.3}$; $AC \approx 3.1$ km.
31. $\tan 62^\circ = \frac{AD}{7.3}$; $AD \approx 13.7$ km.
32. $CD \approx 13.7 - 3.1 \approx 10.6$ km.
33. $\angle PBE = 60^\circ$ because ΔBPE is equilateral.
34. $\angle PBD = m\widehat{PD} = 15^\circ$.
35. BF and BG trisect $\angle ABC$.

TURN OVER

36.

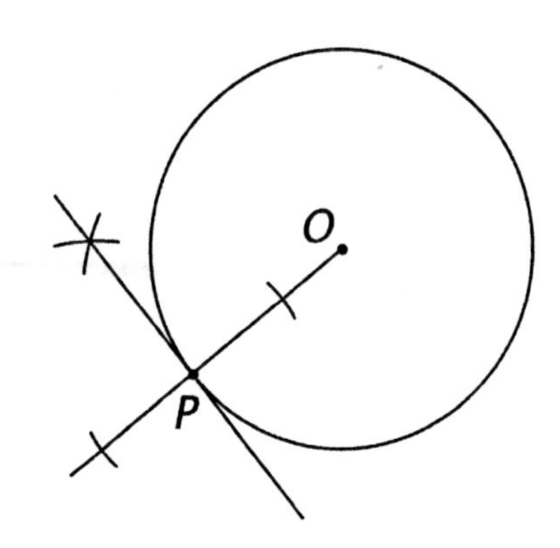

37. $6\pi \approx 18.8$. About 19 inches.

38. $\dfrac{25 \times 12}{19} \approx 16$ turns.

39. 27

40. $0.5A = 27$, so $A = 54$ square ft.

GEOMETRY: *Answer Sheet for* Final Exam–Part 2

Name________________________________

1. *c*
4. *b*
7. *c*
2. *b*
5. *d*
8. *d*
3. *d*
6. *c*

9. *To find the length of a diagonal of a square.*

10. *To find the area of an equilateral triangle.*

11. *To find the perimeter of a regular polygon.*

12. *To find the volume of a sphere.*

13. *a*

14. *b*

15. $(\sin A)^2 + (\cos A)^2 = 1$ *because* $a^2 + b^2 = 1$.

16. $m\overset{\frown}{AD} + m\overset{\frown}{CB} = 90^\circ$.

17. $m\overset{\frown}{DC} = 90^\circ$.

18. $\angle DOC = 90^\circ$.

19. $DO \perp OC$.

20. *Z (or its circumcenter).*

21. *Y (or its centroid).*

22. *Its orthocenter.*

23. *Its circumcenter.*

24. $\sqrt{5}$

25. *AB*, $\frac{1}{2}$; *AC*, 2; *AD*, −2, *AE*, $-\frac{1}{2}$.

26. $AB \perp AD$. *Two lines are perpendicular if the product of their slopes is −1.*

27. *They are parallel.*

28. *Prisms.*

29. $\frac{1}{2}\left(\frac{x}{2}\right)\left(\frac{x}{2}\right)x = \frac{x^3}{8}$

30. $x\left(\frac{x}{2}\sqrt{2}\right) = \frac{\sqrt{2}}{2}x^2$

31.

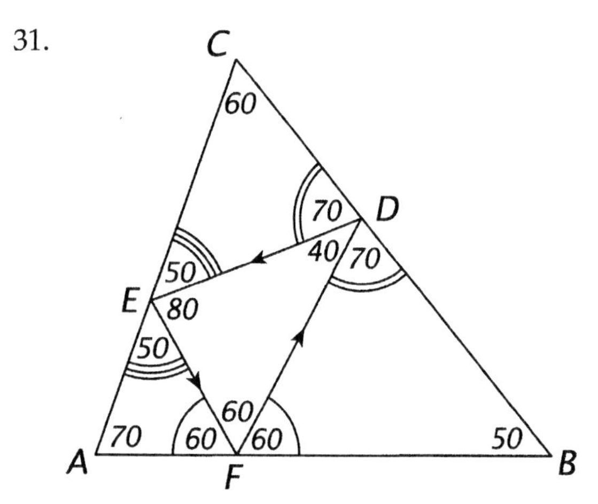

32. *Yes. All of the triangles are similar by AA except ΔDEF.*

33. $\pi r^2 l$

34. $2\pi r l$

35. $\frac{2\pi r l}{\pi r^2 l} = \frac{2}{r}$

TURN OVER

36. *No.*

37. *One with a very small radius.*

38. $\frac{1}{4}2\pi(1) + \frac{1}{4}2\pi(2) + \frac{1}{4}2\pi(3) +$

$\frac{1}{4}2\pi(4) + \frac{1}{4}2\pi(5) + \frac{1}{4}2\pi(6) =$

10.5π.

39. *33 units.*

40. $\frac{1}{4}\pi(3)^2 + \frac{1}{4}\pi(4)^2 + \frac{1}{4}\pi(5)^2 +$

$\frac{1}{4}\pi(6)^2 + 1^2 = 21.5\pi + 1$.